JONES DICTIONARY OF CABLE TELEVISION TERMINOLOGY
3rd Edition

JONES

DICTIONARY

OF

CABLE

TELEVISION

TERMINOLOGY

3rd Edition

: Including
: Related
: Computer &
: Satellite
: Definitions

G L E N N R . J O N E S

™

Published by JONES 21st CENTURY™, INC., a division of
Jones International™, Ltd.
9697 E. Mineral Avenue, Englewood, CO 80112

ISBN: 0-9453-7300-7

■ Epigraph

"The information age is powered by computers and embraces satellites, cable, fiber and other components of high velocity delivery systems. Information and entertainment now move at the speed of light and are like the wind; they know no borders. Time and distance are erased. Information is being delivered faster than it can be understood and there is too much entertainment to watch. So, even in the information age, we must deal with the limitations of the world's first wet computer—the human brain. Therein lies our challenge.

"In the crucible of the marketplace we must embrace converging technologies. We must think not only of how we configure our storage, access and delivery systems; we must contemplate the worth of what we store, access and deliver. We must design around the ultimate destination point of our systems, the human brain. While nurturing profitability, we must bend technology and its consequences to further human needs. We live in a renaissance; in terms of extending the human mind, the world is rich with opportunities."

Glenn R. Jones
March 17, 1987

▪ Acknowledgments

Writing and producing a book of this nature is, of necessity, a group undertaking. For their invaluable assistance, I am grateful to the High Technology Group and to my talented associates at Jones Intercable, Inc., in particular Bob Luff, Group Vice-President/Technology, Ron Hranac, Corporate Engineer, and all the other members of Jones Intercable's Engineering department; Greg Liptak; Jim Carlson; Don Tarkenton; Jim Honiotes; Rick Larson; Lucy Strupp; and Kim Dority. My thanks to all.

Glenn R. Jones
March 1988

■ Dedication

Ever-increasing amounts of new information and entertainment bombard us from every side, every day. The mind simply cannot manage the onslaught by itself.

By utilizing the electronic pipeline called cable television and associated technologies, there is hope that we can attain one of humankind's greatest achievements; the dramatic extension of the human mind.

This dictionary is dedicated to the role that cable can play in that quest.

A See "Ampere."

A-B Switch A high isolation switch used to select between two input signal sources; for example, an off-air antenna and the cable television subscriber drop.

A.C. Nielsen Service bureau that measures audience size for the broadcast and CATV industries.

Aberration An error condition, usually in the cathode-ray tube of a terminal.

Abort To terminate, in a controlled manner, a processing activity in a computer system because it is impossible or undesirable for the activity to proceed.

Abort Timer A device which stops dial-up data transmission if no data are sent within a predetermined time period.

Absolute Address The actual address of a memory location referenced by a computer program, as opposed to the relative or relocatable address. The absolute address of a program or data is usually determined only when the program is loaded into the computer's memory for execution.

Absolute Command In computer graphics, a display command that causes the display device to interpret the data following the command as absolute coordinates, rather than relative coordinates.

Absolute Loader A routine that reads a computer program into main storage, beginning at the assembled origin.

Acceptance Test Actions undertaken to prove that a system fulfills agreed-upon criteria; for example, that the processing of specified input yielded expected results.

Access The manner in which files or data sets are referred to by the computer.

Access Cablecasting Services provided by a cable television system on its public, educational, local government or leased channels.

Access Channels Dedicated channels giving nondiscriminatory access to the cable system by the public, government agencies, or educational institutions.

Access Charge A charge made by the local telephone company for use of the local company's exchange facilities, and/or interconnection with the telecommunications network.

Access Line A telecommunication line that continuously connects a remote station to a data switching exchange (DSE). A telephone number is associated with the access line.

Access Method A technique for moving data between main storage and input/output devices.

Access Mode A technique used to obtain a specific logical record from, or to write a logical record to, a file on a mass storage device; e.g., a disk drive.

Access Point A data element used as a means of entry into a file or record.

Access Time (1) The amount of time required to retrieve the contents of a memory location, determined by the speed of the memory circuitry; (2) The time required by the read/write head of a floppy disk to move from one concentric track of the disk to the next.

Access Trap A method used to detect illegal tampering with the terminal by the subscriber.

Accumulator The primary central processing unit (CPU) register of a computer, and the register to or from which additions and subtractions are usually made.

ACE See "Award for Cablecasting Excellence."

ACL See "Audit Command Language."

Acoustic Coupler A device that allows a conventional telephone handset to feed its signal into a modem, as opposed to direct couplers, which feed the modulated/demodulated signal directly into the phone line.

Acoustic Modem A modulator-demodulator unit that converts data signals to telephone tones and back again.

Active A device or circuit capable of some dynamic function, such as ampli-

fication, oscillation, or signal control, and which usually requires a power supply for its operation.

Active Satellite A satellite that transmits a signal, in contrast to a passive satellite that only reflects a signal. The signal received by the active satellite is usually amplified and translated to a different frequency before it is retransmitted.

Active Tap A CATV feeder device consisting of a directional coupler and a hybrid splitter (e.g., a conventional subscriber tap), in addition to an amplifier circuit.

A/D Conversion See "Analog-to-Digital (A/D) Conversion."

A/D Converter See "Analog-to-Digital (A/D) Converter."

Adapter Mechanism for attaching parts, especially those parts having different physical dimensions or electrical connectors.

Additional Service Television signals that some cable systems may carry in addition to those required or permitted in the mandatory carriage and minimum service categories.

Address A character or group of characters that identifies a register, a particular part of storage, or some other data source or destination.

Addressable The ability to signal from the headend or hub site in such a way that only the desired subscriber's receiving equipment is affected. By this means it is possible to send a signal to a subscriber and effect changes in the subscriber's level of service.

Addressable Register A temporary storage location with a fixed location and address number.

ADI See "Area of Dominant Influence."

Adjacent Channel (1)Any two television channels spaced 6 MHz apart; (2) The channel (frequency band) immediately above or below the channel of interest.

ADP See "Automatic Data Processing."

Advanced Communications Services (ACS) A planned, shared, switched data communications network service that would provide information movement, communications processing, and network management functions. ACS would include the capability of enabling incompatible terminals and computers to communicate with each other. Still in the planning stages, ACS would potentially be as accessible to small data communications users as to large users.

Advanced Television Systems Committee (ATSC) Group sponsored by the Joint Committee on Intersociety Coordination. Purpose is for members of the television and motion picture industry to work together to develop voluntary national standards for advanced television systems.

Advertising Availabilities Time space provided to cable operators by cable programming services during a program for use by the CATV operator; the time is usually sold to local advertisers or used for channel self-promotion.

ADX See "Automatic Data Exchange."

Aerial (1) A device which receives a signal and feeds it in electrical form to a receiver, or which radiates the transmitted electrical signal into space; (2) An antenna; (3) Pertaining to an object positioned above the surface of the earth.

Aerial Cable Outside cable plant that is located on utility poles or other overhead structures.

AFC See "Automatic Frequency Control."

Affiliate Any cable system that receives its programming from a program service.

AFT See "Automatic Fine Tuning."

AGC See "Automatic Gain Control."

Aggregate A transmitted carrier signal that consists of 12 single sidebands being sent over the transmission circuit.

Agile Receiver A satellite receiver that can be tuned to the various transponders available on a communications satellite.

ALC See "Automatic Level Control."

Algebraic Language An algorithmic language of which many statements are structured to resemble the structure of algebraic expressions; for example, ALGOL, FORTRAN.

Algorithm (1) Rule of thumb for doing something with a semblance of intelligence. For example, a descrambling algorithm will yield a clear, unscrambled message from an apparently meaningless one; (2) The procedure used for performing a task.

Alignment The storing of data in relation to certain machine-dependent boundaries. For CATV: adjustment to predefined parameters, conditions or levels.

All-Channel Antenna An antenna which receives signals equally well over a wide band of frequencies. Synonymous with broadband antenna.

All-Number Calling (ANC) The system of telephone numbering that uses all numbers and replaces the two-letter plus five-number $(2L + 5N)$ numbering plan. ANC offers more usable combinations of numbers than the $2L + 5N$ numbering plan and is becoming the nationwide standard.

Allocate To assign a resource, such as a disk or a diskette file, to a specific task.

Allocations Frequency assignment of the communications spectrum by appropriate national or international regulatory authorities.

Alphameric See "Alphanumeric."

Alphanumeric Pertaining to a character set that contains letters, digits, and usually other characters, such as punctuation marks. Synonymous with alphameric.

ALSC See "Automatic Level and Slope Control."

Alternate Channel Interference Interference caused by a signal in the channel beyond an adjacent channel.

Alternating Current (AC) An electrical current, the polarity of which is periodically reversed. In each complete polarity reversal, or cycle, the alternation starts at zero, rises to a maximum positive level, returns to zero, continues to a maximum negative level, and again returns to zero. The frequency of the alternating current is the number of complete cycles each second.

ALU See "Arithmetic Logic Unit."

Aluminum Sheath Cable Coaxial cable constructed with a solid copper or copper-clad aluminum center conductor, dielectric insulation, and semi-rigid aluminum outer conductor.

AM See "Amplitude Modulation."

Ambient Temperature The temperature surrounding apparatus and equipment. Synonymous with room temperature.

American National Standards Institute (ANSI) Established in 1918 for the purpose of establishing nationally coordinated voluntary safety, engineering, and industrial standards. Members include industrial firms, trade organizations, technical societies, labor and consumer organizations, and government agencies.

American Standard Code for Information Interchange (ASCII) A character set that includes the upper case and lower case English alphabet, numerals, special symbols, and 32 control codes. Each character is represented by a seven-bit binary number. Therefore, one ASCII-encoded character can be stored in one byte of computer memory.

AML See "Amplitude Modulated Link."

Ampere (A) A unit of measure for electrical current equivalent to a flow of one coulomb per second, or to the steady current produced by one volt applied across a resistance of one ohm.

Amplifier Device used to increase the operating level of an input signal. Used in a cable system's distribution plant to compensate for the effects of attenuation caused by coaxial cable and passive device losses.

Amplifier Spacing The physical or electrical distance between two amplifiers, frequently expressed electrically in decibels.

Amplify To boost the signal levels.

Amplitude The size or magnitude of a voltage or current waveform; the strength of a signal.

Amplitude Modulated Link (AML) A form of microwave communications using amplitude modulation for the transmission of television and related signals.

Amplitude Modulation (AM) The form of modulation in which the amplitude of the signal is varied in accordance with the instantaneous value of the modulating signal.

Analog Pertaining to signals in the form of continuously variable physical quantities.

Analog Channel A communication channel on which the information transmitted can take any value between the limits defined by the channel. Voice-grade channels are analog channels.

Analog Computer (1) A computer in which analog representation of data is mainly used; (2) A computer that operates on analog data by performing physical processes of these data.

Analog Data Data represented by physical quality considered to be continuously variable and whose magnitude is made directly proportional to the data or to a suitable function of the data.

Analog Facility Terminal (AFT) A voice-frequency facility terminal that performs signaling and transmission functions and includes analog channel banks. It interfaces between an analog carrier system and a switching system, a metallic facility, a digital terminal, or another analog facility terminal.

Analog-to-Digital (A/D) Conversion The conversion of an analog signal into a digital equivalent. An A/D converter samples or measures an input voltage and outputs a digitally encoded number corresponding to that voltage.

Analog-to-Digital (A/D) Converter (1) A functional unit that converts analog signals to digital data; (2) A device that senses an analog signal and converts it to a proportional representation in digital form; (3) An electromechanical device that senses an electrical signal and converts it to a proportional representation in digital form.

Analog Transmission Transmission of a continuously variable signal as opposed to a discrete signal. Physical quantities such as temperature are described as analog while data characters are coded in discrete pulses and are referred to as digital.

ANC See "All-Number Calling."

Anchor The buried anchor and eyerod to which the utility pole guy-wire is attached.

Anchor Rod The metal rod that attaches the anchor to the strand guy wire.

ANI See "Automatic Number Identification."

Anisochronous Transmission A transmission process in which there is always an integral number of unit intervals between any two significant instants in the same group.

ANSI See "American National Standards Institute."

Answerback The response of a terminal to remote control signals.

Antenna Any structure or device used to collect or radiate electromagnetic waves.

Antenna Array A radiating or receiving system composed of several spaced radiators or elements.

Antenna Height Above Average Terrain The average of the antenna height above the terrain from two to ten miles from the center of the antenna for the eight directions spaced evenly for each 45 degrees of azimuth starting with true North. Where circular or elliptical polarization is used, the antenna height above average terrain is based upon the height of the radiation center of the antenna which transmits the horizontal component of radiation.

Antenna Power The product of the square of the broadcast antenna current and the antenna resistance where the current is measured.

Antenna Power Gain The ratio of the power required to produce a selected field strength with an antenna of interest, to the power required for the same field strength with a reference, dipole, or isotropic antenna (at the same test location).

Antenna Preamplifier A small signal booster located in the immediate vicinity of the antenna, used to amplify extremely weak signals, thereby improving the signal-to-noise ratio of a system.

Antenna Resistance (1) The total resistance of the antenna system at the operating frequency and at the point at which the antenna current is measured; (2) That part of the antenna impedance which is resistive.

Antenna Run Transmission lines which extend from the receiving antenna to the headend or the beginning of the distribution lines.

Antenna Thrust Weight, in pounds per square inch, exerted on antenna support members under various wind and icing conditions.

Antiope The French system for character coding and display on viewdata and teletext terminals.

Anti-Siphoning Rules Federal Communications Commission rules that prohibit cable systems from televising programs on pay cable channels that might otherwise be offered on broadcast television channels. The anti-siphoning rules restrict cable systems to showing movies no older than three years and sports events not normally offered on conventional broadcast television.

Apartment Box A locked, protective enclosure used at apartment and other multidwelling complexes to house CATV active and passive devices.

Aperture One or more adjacent characters in a mask that cause retention of the corresponding characters.

APL See "Average Picture Level."

Application The use to which a data processing system is put; for example, a payroll application, an airline reservation application, a network application.

Application Software In computers, programs used to interact with and accomplish work for the user. Application software is usually written in a "higher language," such as BASIC, COBOL, FORTRAN, or PASCAL, and may be written by the user or supplied by the manufacturer or a software company.

Applications Program A computer program dedicated to a specific purpose or task. Applications programs, which produce discernible results and can sometimes be machine-independent, are distinct from systems programs, which are designed to drive particular electronic devices and are always machine-dependent.

Arbitron A service organization that measures and reports audience viewership data related to programming to broadcasters and cable television operators.

Architecture (1)The internal configuration of a processor (including its registers and instruction set) or network; (2) In CATV, the type of distribution network used; for example, tree/branch.

Archive (1) The process of storing data files in a retrievable form; (2) The data files so stored.

Area Code A three-digit number identifying one of 152 geographic areas of the United States and Canada to permit direct distance dialing on the telephone system.

Area of Dominant Influence (ADI) Geographic area where a broadcast signal measures at or above a predetermined level.

Arithmetic Logic Unit (ALU) The portion of the central processor that executes all mathematical and logical functions.

Armored Cable Coaxial cable that can be direct buried without protective conduit, or used in underwater applications. This type of cable is constructed with a flooding compound applied to the cable's outer shield, followed by plastic jacketing, steel armor and flooding compound, and an additional plastic jacket.

Array An arrangement of elements in one or more dimensions.

Artificial Intelligence The capability of a device to perform functions that are normally associated with human intelligence, such as reasoning, learning, and self-improvement.

ARU See "Audio Response Unit."

ASA American Standards Association; former name of the American National Standards Institute.

ASC See "Automatic Slope Control."

ASCII See "American Standard Code for Information Interchange."

ASCII Keyboard A keyboard including keys for all of the characters of the ASCII character set. Usually includes three cases for each alpha character: uppercase, lower case and control.

ASGC See "Automatic Slope and Gain Control."

Aspect Ratio The ratio of picture width to height (4 to 3 for North American NTSC broadcast video).

Assemble To translate a program expressed in an assembly language into a computer language and perhaps to link subroutines. Assembling is usually accomplished by substituting the computer language operation code for the assembly-language operation code and by substituting absolute addresses, immediate addresses, relocatable addresses, or virtual addresses for symbolic addresses.

Assembler A computer program used to assemble. Synonymous with assembly program.

Assembly Language A low-level programming language that permits the programmer to use mnemonics for function codes.

Assembly Program See "Assembler."

Associated Broadcasting Station The broadcasting station with which a remote pickup broadcast base or mobile station is licensed as an auxiliary and with which it is principally used.

Asynchronous (1) Having a variable time interval between successive bits, characters, or events. In asynchronous data transmission, each character is individually arranged, usually by using start and stop bits; (2) Descriptive of the transmission method, or the terminal equipment employed, which is self-clocking.

Asynchronous Time-Division Multiplexing An asynchronous signal transmission mode that makes use of time-division multiplexing.

ATSC See "Advanced Television Systems Committee."

Attended Operation Operation of a station by a qualified operator on duty at the place where the transmitting apparatus is located with the transmitter in plain view of the operator.

Attenuation The difference between transmitted and received power due to loss through equipment, lines, or other transmission devices; usually expressed in decibels.

Attenuator A device for reducing the amplitude of a signal.

Audio Relating to sound or its reproduction; used in the transmission or reception of sound.

Audio Cassette A small cartridge having magnetic recording tape mounted on rotatable reels. Its size is standardized, allowing interchange between a wide variety of record/playback equipment.

Audio Channel A channel capable of satisfactorily transmitting signals within the audio range.

Audio Frequency A frequency lying within the audible spectrum (the band of frequencies extending from about 20 Hz to 20 KHz).

Audio Programmer Device that permits automatic programming of music and announcement tape cartridges to coincide with slide projection.

Audio Response A form of output that uses verbal replies to inquiries. The computer is programmed to seek answers to inquiries made on a time-shared on-line system and then to utilize a special audio response unit which elicits the appropriate prerecorded response to the inquiry.

Audio Response Unit An output device that provides a spoken response to digital inquiries from a telephone or other device. The response is composed from a prerecorded vocabulary of words and can be transmitted over telecommunication lines to the location from which the inquiry originated.

Audio Track Geometry The tape location and width of the audio tracks with their associated guard bands.

Audit Command Language (ACL) A high-level programming language.

Aural Broadcast Intercity Relay Station A fixed station utilizing radiotelephony for retransmission of aural program material between broadcast stations, for simultaneous or delayed broadcast.

Aural Broadcast STL Station A fixed station utilizing radio-telephony for the transmission of aural program material between a studio and the transmitter of a broadcasting station for simultaneous or delayed broadcast.

Aural Cable Services providing FM-only original programming to cable systems on a lease basis.

Aural Carrier The sound portion of a television signal.

Aural Center Frequency (1)The average frequency of an emitted signal when modulated by an aural (audio) signal; (2) The frequency of the emitted wave without modulation. Usually refers to frequency modulation methods.

Aural Transmitter The broadcast equipment used for the transmission of the sound portion of a television channel.

Authorization The right granted to a user to communicate with or make use of a computer system.

Authorization Code A code, made up of a user's identification and password, used to protect against unauthorized access to data and system facilities.

Autodial Having the ability to automatically dial a pre-designated telephone number or numbers.

Automated Insertion Technology that enables a cable operator to automatically utilize advertising availabilities made available by program providers to insert commercial announcements into a particular program.

Automatic Call Distributor A system for automatically providing even distribution of incoming calls to operator or attendant positions. Calls are served in the approximate order of arrival and are routed to positions in the order of their availability for handling a call.

Automatic Calling Unit A dialing device that permits a business machine to automatically dial calls over a network.

Automatic Data Exchange (ADX) An automatic exchange in a data transmission network.

Automatic Data Processing (ADP) (1) Data processing performed by computer systems; (2) Data processing largely performed by automatic means; (3) The branch of science and technology concerned with methods and techniques relating to data processing largely performed by automatic means; (4) Pertaining to data-processing equipment such as electrical accounting machines and electronic data-processing equipment; (5) Data processing by means of one or more devices that (a) use common storage for all or part of a program and also for all or part of the data necessary for execution of the program, (b) execute user-written or user-designated programs, (c) perform user-designated symbol manipulation such as arithmetic operations, logic operations, or character-string manipulations, and (d) execute programs that can modify themselves during their execution.

Automatic Dialing Unit A device capable of automatically generating dialing digits.

Automatic Fine Tuning (AFT) An electronic circuit which automatically tracks a given frequency through small positive and negative drifts, maintaining maximum signal strength reception.

Automatic Frequency Control (AFC) A system which keeps a circuit automatically tuned to a desired signal frequency.

Automatic Gain Control (AGC) A circuit which automatically controls the operating level of an amplifier so that the output signal remains relatively constant despite varying input signal level.

Automatic Intercept System A type of traffic service system consisting of one or more automatic intercept centers and a centralized intercept bureau for handling intercept calls.

Automatic Level and Slope Control (ALSC) A circuit that automatically keeps the gain (level) and tilt (slope) response of a CATV amplifier adjusted to compensate for variations in input signal levels.

Automatic Level Control (ALC) The automatic adjusting of signal levels in a system (similar to automatic gain control).

Automatic Mobile Relay Station A remote pick-up broadcast base station actuated by automatic means and used to relay communications between base and mobile stations, and from mobile stations to broadcast stations.

Automatic Number Identification (ANI) The automatic identification of a calling station, usually for automatic message accounting. Also used in pay-per-view automated telephone order entry to identify a customer for billing and program authorization purposes.

Automatic Request for Repetition A feature that automatically initiates request for retransmission when an error in transmission is detected.

Automatic Send/Receive (ASR) A teletypewriter unit with keyboard, printer, paper tape, reader/transmitter, and paper tape punch. This combination of units may be used on-line or off-line and, in some cases, on-line and off-line concurrently.

Automatic Slope Control (ASC) A circuit in a CATV amplifier that automatically keeps the tilt (slope) response adjusted to compensate for cable attenuation changes due to temperature variations.

Automatic Slope and Gain Control (ASGC) A circuit in a CATV amplifier that combines the functions of automatic gain and automatic slope control circuits.

Automatic Spacing A method whereby unavoidable errors in amplifier spacing are automatically corrected.

Automatic Temperature Control A method whereby attenuation changes in amplifiers or coaxial cable due to ambient temperature variations are automatically corrected.

Automatic Tilt Automatic correction of changes in tilt, or the relative level of signals of different frequencies. See also "Tilt" and "Automatic Slope Control."

Automatic Volume Control (AVC) A system that holds the gain, and subsequently the output, of an audio circuit relatively constant, despite input signal amplitude variations.

Auxiliary Eye A second anchor eye attached to an existing anchor rod for installation of an additional guy.

Auxiliary Transmitter A transmitter maintained only for transmitting of the regular programs of a station in case of a failure of the main transmitter.

AVC See "Automatic Volume Control."

Average Daily Circulation The estimated number of different households reached by a particular newspaper or station on each day of the week.

Average Picture Level (APL) The average level of the picture signal during active scanning time integrated over a frame period and expressed as a percentage of the blanking to reference white range.

Award for Cablecasting Excellence (ACE) A series of awards for original made-for-cable programs.

B

Babble Undesired and unintelligible signals inadvertently imposed on a desired audio signal.

Back Light Directional illumination coming mostly from behind the subject.

Back Matched Tap A cable tap device which employs transformer isolation and also employs impedance matching at the tap-off points.

Back-Mounted A connector attached from the inside of a box, having mounting flanges placed on the inside of the machines.

Back Porch That portion of the composite video signal which lies just after the trailing edge of the horizontal sync pulse.

Backfill Soil, rocks, and other material used to fill up a trench or hole.

Background Display Image See "Static Display Image."

Background Ink In optical character recognition, a type of ink with high reflective characteristics that is not detected by the scan head, and thus is used for print location guides, logotypes, instructions, and any other desired preprinting that would otherwise interfere with reading.

Background Light A separate illumination of the background or set.

Background Noise (1) Extra bits or words ignored or removed from the data at the time it is used; (2) Errors introduced into the data in a system; (3) Any disturbance tending to interfere with the normal operation of a system or unit.

Backup Copy A copy of a file or data set that is kept for reference in case the original file or data set is destroyed.

Balance (1) To distribute traffic over the line terminals at a central office as uniformly as possible. Without load balancing, a portion of the switching equipment may become overloaded even though the total capacity of the system has not been exceeded; (2) To adjust the impedance of circuits and balance networks to achieve specified return loss objectives at junctions of two-wire and four-wire circuits; (3) To adjust amplifier operating levels in a cable television distribution plant.

Band (1) A group of tracks on a magnetic

drum or on one side of a magnetic disk; (2) The frequency spectrum between two defined limits.

Banding One or more groups of four bands in each reproduced field containing a different video level and/or signal-to-noise ratio, as compared to the rest of the picture or to other groups of bands.

Bandpass Filter A device which allows signal passage to frequencies within its design range and which effectively bars passage to all signals outside that frequency range.

Bandwidth (1) A measure of the information-carrying capacity of a communication channel. The bandwidth corresponds to the difference between the lowest and highest frequency signal which can be carried by the channel; (2) The range of usable frequencies that can be carried by a cable television system.

Bar-Code Scanner An optical device that reads data from documents having characters recorded in the form of parallel bars. The characters are translated into digital signals for storage or processing.

Barndoor Flaps or shutters attached to the front of a luminaire.

Base Light An extremely diffused, over-all illumination in the studio coming from no apparent source.

Baseband The band of frequencies occupied by the signal in a carrier wire or radio transmission system before it modulates the carrier frequency to form the transmitted line or radio signal.

Baseband Channel Connotes that modulation is used in the structure of the channel, as in a carrier system. The usual consequence is phase or frequency offset. The simplest example is a pair of wires that transmits direct current and has no impairments such as phase offset or frequency offset that would destroy waveform.

BASIC (Beginner's All-Purpose Symbolic Instruction Code) A computer programming language.

Basic Cable Service The service that cable subscribers receive for the threshold fee, usually including local television stations, some distant signals and perhaps one or more non-broadcast services.

Batch (1) An accumulation of data to be processed; (2) A group of records or data-processing jobs brought together for processing or transmission.

Batched Communication The sending of a large body of data from one station to another station network, without intervening responses from the receiving unit.

Baud A measure of signaling rate.

Baud Rate The transmission rate that is, in effect, synonymous with signal events, usually bits per second. Baud rate may be equal to or greater than the bit rate, since one bit can be made up of more than one signal event.

BCD Code See "Binary-Coded Decimal Code."

Beam Angle See "Beamwidth."

Beam Deflection On a CRT display device, the process of changing the orientation of the electron beam.

Beamwidth The angular extent over which an antenna detects or transmits at least 50 percent of its maximum power.

Beats The unwanted sum and/or difference frequencies resulting from the heterodyning (mixing) of two or more signals.

Beyond the Horizon Region That physical region beyond the optical horizon with which line-of-sight radio communications is not normally possible, but can occur if atmospheric conditions are such to cause beam bending or forward scattering of the radio signal.

Bicycle (tapes) The process whereby video tape material is distributed by sending or "bicycling" the tape after presentation to the next site for its scheduled presentation.

Bidirectional A pathway allocating two-way data or communication exchange.

Bidirectional Flow Flow in either direction represented on the same flow line in a flowchart.

Bildchirmtext The West German post, telegraphs, and telephones' (PTT's) view system.

Binary The base-two number system; it uses only the digits 0 and 1.

Binary-Coded Decimal (BCD) Code Method of transmitting binary data that utilizes fewer wires in the transmission cable than do alternate methods.

Binary Digit In binary notation, either of the characters 0 or 1. Frequently shortened to "bit."

Bird See "Communications Satellite."

Bit (1) A contraction of the words "binary digit," the smallest unit of information; (2) A single character in a binary number; (3) A single pulse in a group of pulses; (4) A unit of information capacity of a storage device.

Bit Density A measure of the number of bits received per unit of length or area.

Bit Rate The speed at which bits are transmitted, usually expressed in bits per second. See also "Baud."

Bits Per Second (BPS) (B/S) Digitial information rate expressed as the number of binary information units transmitted per second.

Black Clipper A piece of equipment or a circuit which does not transmit black peak below a certain pre-set level of picture signal and at the same time transmits the remainder of the input signal without change.

Black Compression The compression of the steps toward the black end of a staircase wave-form or gray scale. The reduction of contrast in the dark gray to black range of a television picture.

Black Level That level of picture signal corresponding to the maximum limit of black peaks.

Black Peak The maximum excursion of the picture signal in the black direction during the time of observation.

Blacker-Than-Black The amplitude region of the composite video signal below reference black level in the direction of the synchronizing pulses.

Blanket Area The area within which the radio signal is at least 1 volt per meter—an unusually strong signal.

Blanket Contour The boundary of the blanket area.

Blanketing A form of interference which is caused by the presence of an interfering signal of great intensity, usually 1 volt per meter or greater.

Blanking (picture) The portion of the composite video signal whose instantaneous amplitude makes the vertical and horizontal retrace invisible.

Blanking Level The level of the front and back porches of the composite video signal.

Blanking Pulse (1) A signal used to cut off the electron beam and thus remove the spot of light on the face of a television picture tube or image tube; (2) A signal used to suppress the picture signal at a given time for a required period.

Blanking Signal A specified series of blanking pulses.

BLCT See "Broadcast License Commercial Television."

Bleeding Whites An overloading condition in which white areas appear to flow irregularly into black areas.

Block A group of bits, or characters, transmitted as a unit. An encoding procedure is generally applied to the group of bits or characters for error control purposes.

Block Converter An electronic device that changes a group of frequencies to a lower or higher group of frequencies. In satellite communications, a block converter can be used to change the band of received satellite signals from microwave frequencies (3.7-4.2 GHz) to UHF frequencies (950-1450 MHz). In CATV, a block converter may be used to change one group of cable channels to another group of channels compatible with the customer's television set.

Block-Error Rate The ratio of the number of blocks incorrectly received to the total number of blocks sent.

Block Tilt An approximation of linear tilt (slope) achieved by operating signal levels in groups or blocks of various flat amplitudes. The higher frequency blocks are generally operated at higher successive levels than lower frequency blocks.

Blooming An unwanted, distorted increase in picture image size and defocusing, usually caused by improper high voltage or focus CRT circuits.

Board See "Printed Circuit Board."

Body Belt A belt worn by linemen who climb utility poles. The belt has loops for tools, and attachment rings for the safety strap that goes around the pole.

Bonding (1) The permanent joining of metallic parts to form an electrically conductive path which will ensure electrical continuity and the capacity to conduct safely any current likely to be imposed; (2) The interconnection of the CATV cable support strand with a telephone company support strand and/or the power company neutral / ground wire to eliminate ground potential differences.

Bootstrap A short loader program that loads a more sophisticated loader into memory. That loader, in turn, loads the desired program. The term "bootstrap" arises from the idea that the computer is picking itself up by its bootstraps. In other words, it progresses from the bootstrap to the loader to the main program itself.

Bounce (1) Unnatural sudden variation in the brightness of a television picture. Sudden undesired change in video level as shown by a wave-form monitor; (2) A test signal used to simulate rapid video level changes and gauge a television system's response to bounce; (3) Unwanted multiple vibrations in switch or key closures causing a stuttering effect of the controlled function.

BPO See "British Post Office."

BPS See "Bits Per Second."

Branch Cable A cable that diverges from a main cable to reach some secondary point.

Branching A computer operation, such as switching, where a choice is made between two or more possible courses of action depending upon some related fact or condition.

Branching Networks Electrical networks, such as filters, isolators, and circulators, used for transmission or reception of signals over two or more channels on one antenna.

Breadboard An experimental model of a unit used to test the operation of the design.

Breathing Unnatural variation at a slow regular rate in the brightness of a television picture. Slow regular variation in video level as shown by a wave-form monitor.

Breezeway That portion of the back porch between the trailing edge of the sync pulse and the start of the color burst.

Bridger See "Bridging Amplifier."

Bridging Amplifier An amplifier connected directly into the main trunk of the CATV system. It serves as a sophisticated tap, providing isolation from the main trunk, and has multiple high level outputs that provide signal to the feeder portion of the distribution network. Synonymous with bridger and distribution amplifier.

Brightness The quantity or intensity of light given out perpendicular to the surface emitting it, per unit area of surface.

British Post Office (BPO) The British post, telegraphs, and telephones (PTT) system.

Broadband Any system able to deliver multiple channels and/or services to its users or subscribers. Generally refers to CATV systems. Synonymous with wideband.

Broadcast License Commercial Television (BLCT) A television broadcasting station licensed to include commercial advertisements in its programming as a means of generating additional income.

Broadcast Television Systems Committee (BTSC) An Electronics Industries Association (EIA) committee created to develop a standardized approach to the implementation of the multichannel television sound (MTS). Following laboratory tests of several potential MTS systems, the one selected was a combination using the transmission parameters developed by Zenith Electronics Corp. and the noise reduction system developed by dbx, Inc.

Broadcaster's Service Area The geographic area that receives a station's transmission signal.

Broadcasting (1) The dissemination of any form of radio electric communications by means of Hertzian waves intended to be received by the public; (2) Transmission of over-the-air signals for public use. See also "Point-to-Point"

and "Point-to-Multipoint."

Broadcasting Authority-Receive The broadcast authority at the receiving end of an international sound program or television connection. Synonymous with Television Authority-Receive.

Broadcasting Authority-Send The broadcast authority at the sending end of an international sound program or television connection. Synonymous with Television Authority-Send.

Broadside Array An antenna array designed to provide maximum signal radiation in the direction broadside, or perpendicular, to the array.

BTSC (1) See "Broadcast Television Systems Committee"; (2) The defacto multichannel television sound (MTS) standard developed by dbx, Inc. and Zenith for stereo audio television broadcasting, comprised of left plus right modulating the aural carrier in the conventional way (for compatibility with existing monaural television sets), a stereo pilot at 15.734 KHz (1H), dbx companded left-right double sideband suppressed subcarrier at 31.468 KHz (2H), a monaural FM SAP channel subcarrier at 78.67 KHz (5H), and a professional (PRO) channel FM subcarrier (for telemetry, etc.) at 102.271 KHz (6.5H).

Bucket Truck Truck-mounted extension boom for above-ground maintenance. Synonymous with cherry picker.

Buffer Memory area in computer or peripheral device used for temporary storage of information that has just been received. The information is held in the buffer until the computer or device is ready to process it. Hence, a computer or device with memory designated as a buffer area can process one set of data while more sets are arriving.

Bug (1) A system or programming problem. Also refers to the cause of any hardware or software malfunction. May be random or non-random; (2) A special hand-operated telegrapher's key used to send Morse or other codes at high speeds.

Building Circuit A circuit located completely within one building.

Bulk Storage Out-of-use term for mass storage.

Bulk Transfer The transfer of a batch of data in a continuous burst by direct link between computers, or by magnetic tape transfer.

Bundespost The West German post, telegraphs, and telephones (PTT) system.

Buried Cable Outside cable plant located underground.

Burn-In The phase of component testing where basic flaws or early failures are screened out by running the circuit for a specified length of time, such as a week, generally at increased temperatures in some sort of environmental test chamber.

Burning The process of programming a read-only memory (ROM).

Burst (1) In data communication, a sequence of signals counted as one unit in accordance with some specific criterion or measure; (2) A color burst.

Burst Modem In satellite communications, an electronic device used at each station that sends high-speed bursts of data which are interleaved with one another. These bursts must be precisely timed to avoid data collisions with multiple stations.

Burst Transmission Data transmission at a specific data signaling rate during controlled intermittent intervals.

Bus A circuit or group of circuits which provide an electronic pathway between two or more central processing units (CPUs) or input/output devices.

Bus Controller The unit in charge of generating bus commands and control signals.

Bussback The connection, by a common carrier, of a circuit back to the input portion of a circuit.

Byte A group of bits treated as a unit used to represent a character in some coding systems. The values of the bits can be varied to form as many as 256 permutations. Hence, one byte of memory can represent an integer from 0 to 255 or from -127 to $+128$.

C

C-Band (1) The group of microwave frequencies from 4 to 6 GHz; (2) The band of satellite downlink frequencies between 3.7 and 4.2 GHz, which are also shared with terrestrial line-of-sight microwave users.

C-MOS See "Complementary Metal-Oxide Semiconductor."

C-Span Cable Satellite Public Affairs Network. A satellite-delivered service that provides coverage of the U.S. House of Representatives and other governmental affairs programming. C-Span II provides coverage of the U.S. Senate.

CA See "Commercial Announcement."

CAB See "Cable Advertising Bureau."

Cable (1) See "Cable Television"; (2) One or more electrical or optical conductors found within a protective sheathing. When multiple conductors exist, they are isolated from each other.

Cable Advertising Bureau (CAB) Trade association for the cable television industry primarily devoted to promotion of advertising sales on local, regional, and national levels.

Cable Communications Policy Act of 1984 This act, passed by Congress in 1984, updates the original Communications Act of 1934. The primary changes dealt with cable television regulation, theft of service, equal employment opportunity (EEO) and various licensing procedure changes.

Cable Compatible Generally refers to consumer devices, such as television sets and videocassette recorders, that are designed and constructed to allow direct connection of a CATV subscriber drop to the device. Frequently, they have a tuner capable of receiving cable channels other than 2-13 (e.g., midband, superband, and hyperband channels). Even though a device may be cable compatible, it may still require an external descrambler to receive scrambled channels.

Cable Friendly Technology that is compatible with other CATV products or programs.

Cable Powering A method of supplying electrical power through the coaxial cable to system amplifiers.

Cable Ready See "Cable Compatible."

Cable Spacer A device used in lashed cable construction to provide a separation between the cable and the support strand.

Cable Support Strap A supporting device used over the messenger strand to hold cable and cable spacers in position.

Cable Television A broadband communications technology in which multiple television channels as well as audio and data signals are transmitted either one way or bidirectionally through a distribution system to single or multiple specified locations. The term also encompasses the associated and evolving programming and information resources that have been and are being developed at the local, regional, and national levels.

Cable Television Relay Station (CARS) A fixed or mobile microwave communications station used for the transmission of television and related audio signals, FM broadcast stations, cablecasting, data or other information, or test signals for reception at one or more fixed receive points from which the signals are then distributed to the public by cable.

Cable Television Relay Studio-to-Head-end Link (SHL) Station A community antenna relay (CAR) transmitter licensed by the FCC for use on frequencies in which fixed operations are permitted for the transmission of television signals back to the headend or studio location. Most often used in news gathering or sporting events.

Cable Television System (CATV) A broadband communications system capable of delivering multiple channels of entertainment programming and non-entertainment information from a set of centralized antennas, generally by coaxial cable, to a community. Many cable television designs integrate microwave and satellite links into their overall design, and some now include optical fibers as well. Previously called "Community Antenna Television."

Cablecasting Origination of programming, usually other than automated alphanumeric services, by a CATV system.

Cache In a processing unit, a high-speed buffer storage that is continually updated to contain recently accessed contents of main storage. Its purpose is to reduce access time.

CACS See "Classified Ad Channel System."

CAFM Cable FM.

CAI See "Computer-Assisted Instruction."

Call Sign Station identification consisting of a combination of letters and, sometimes, numerals, required by broadcasting regulation.

Camera Control Unit (CCU) An electronic device that provides all the operating voltages and signals for the proper set up, adjustment and operation of a television camera.

Candle Power A measure of intensity of a light source in a specific direction.

Capacitance A measure of the ability to store electric charge.

Capacitive Tap A now-obsolete pressure tap with a capacitor network providing the desired amount of loss and isolation between the feeder cable and the subscriber drop cable.

Capacitor Storage A storage device that uses the capacitive properties of certain materials.

CAPTAINS See "Character and Pattern Telephone Access Information Network Systems."

Carriage The carrying of certain television station signals on the cable system's channels. The Federal Communications Commission specifies which channels must be carried.

Carrier An electromagnetic wave of which some characteristic is varied in order to convey information.

Carrier-to-Noise Ratio In cable television, the ratio of peak carrier power to root mean square (RMS) noise power in a 4 MHz bandwidth.

Carrier Wave An electromagnetic wave whose amplitude, frequency, or phase can be modulated to transmit information such as a television signal.

CARS See "Cable Television Relay Station," "Community Antenna Relay Service," and "Community Antenna Relay Station."

Cascadability The performance capability of a cascade of amplifiers used to reamplify the same signal along a cable system.

Cascade The operation of two or more devices (such as amplifiers in a cable television system) in series so that the output of one device feeds the input of the next.

Cash Flow See "Operating Cash Flow."

Cassette A holder (magazine) of reels of magnetic tape or film that is itself a mechanical subassembly, which may be easily inserted into and removed from a record or playback machine.

CAT See "Computer-Assisted Training."

CATA See "Community Antenna Television Association."

Cathode-Ray Oscilloscope (CRO) Electronic measuring device that utilizes a cathode-ray tube as a display device. See also "Oscilloscope."

Cathode-Ray Tube (CRT) (1) A vacuum tube display in which a beam of electrons can be controlled to form alphanumeric characters or symbols on a luminescent screen; (2) The picture tube in a television set.

CATV Community antenna television. See "Cable Television System."

CCD See "Charge-Coupled Device."

CCETT See "Centre Commun d'Etudes de Television et de Telecommunications."

CCIR Comité Consultatif International des Radio Communications. See "International Consultative Committee for Radio."

CCITT Comité Consultatif International de Telegrafique et Telephonique. See "International Telegraph and Telephone Consultative Committee."

CCTV See "Closed-Circuit Television."

CCU See "Camera Control Unit."

Center Frequency (1) The average frequency of the emitted wave when modulated by a sinusoidal wave; (2) The frequency of the emitted wave without modulation.

Central Computer In data transmission, the computer that lies at the center of a network and generally does the basic centralized functions for which the network was designed. Synonymous with host computer and host processor. Multiple central computers or hosts are sometimes also configured to work together in a larger network.

Central Processing Unit (CPU) The unit of a computer that includes circuits controlling the interpretation and execution of instructions.

Centralized (Computer) Network A computer configuration in which one computer or a group of co-located computers provides computing power and maintains control of applications programs, data, and telecommunications, sometimes over a variety of other computers and/or peripheral equipment, such as printers.

Centralized Processing Computer operations in which one computer or a group of co-located computers provide computing services and maintain network control.

Centre Commun d'Etudes de Television et de Telecommunications (CCETT) The French post, telegraphs, and telephones research center.

Centrex A telephone PABX equipment service that allows dialing within the system, direct inward dialing, and automatic identification of outward dialing, and that can be used to limit outward long distance dialing.

CEPT See "European Conference of Postal and Telecommunications Administrations."

Certificate of Compliance Authorization issued by the FCC for the operation of a cable television system in a community or for the carriage of additional television signals by an operating cable television system.

Chaining The process of having one program transfer control to another program.

Channel A signal path of specified bandwidth for conveying information.

Channel Capacity In a cable television system, the number of channels that can be simultaneously carried on the system. Generally defined in terms of the number of 6 MHz (television bandwidth) channels.

Channel Combiner An electronic or passive device which accepts the radio frequency (RF) signals from many sources and combines them for transmission on the cable. See also "Combining Network."

Channel Frequency Response (1) The relationship within a cable television channel between amplitude and frequency of a constant amplitude input signal as measured at a subscriber terminal; (2) The measure of amplitude-frequency distortion within a specified bandwidth.

Channel Mode The terminal mode for channel selection, entered by pressing the "Channel" button.

Character (1) One of the symbols in a code; (2) In computers, a digit, letter, or symbol used alone or in some combination to express information, data or instructions.

Character and Pattern Telephone Access Information Network Systems (CAPTAINS) A viewdata system used in Japan.

Character Generator An alphanumeric text generator, commonly used to display messages on a television set. Some sophisticated versions also include color, graphics, and mass memory for text storage.

Charge-Coupled Device (CCD) A solid-state device used in many television cameras to convert optical images into electronic signals. These imagers are organized into rows and columns of image elements called pixels. The charge pattern formed in the CCD pixels when light strikes them forms the electronic representation of the image.

Cherry Picking Selection of programming from various sources.

Chip (1) In micrographics, a piece of microform that contains both micro-images and coded identification; (2) A minute piece of semiconductive material used in the manufacture of electronic components; (3) An integrated circuit.

Chrominance Signal The color signal component in color television that represents the hue and saturation levels of the colors in the picture.

Churn (1) Subscriber activity relating to disconnections; (2) Upgrading, downgrading or otherwise changing levels of service.

Circuit (1) In communication systems, an electronic, electrical, or electromagnetic path between two or more points capable of providing a number of channels; (2) Electric or electronic part; (3) Optical or electrical component that serves a specific function or functions.

Circuit Board A board to which is affixed electronic circuitry or components.

Circuit Breaker A manual or automatic protective device for closing and opening a circuit between separate contacts under both normal and abnormal conditions.

Circuit Section Part of a telecommunications circuit, the terminals of which are accessible at baseband frequencies.

Clamper A device which functions during the horizontal blanking or sync interval to fix the level of the picture signal at some predetermined reference level at the beginning of each scanning line.

Clamping The process of re-establishing the direct current (DC) reference level of the picture signal at the beginning of each scanning line.

Class I Cable Television Channel A cable television channel whose source is a television broadcast signal that is being presently transmitted to the public and conveyed to the cable system for retransmission to the public, direct connection, off-the-air or obtained indirectly by microwave or by direct connection to a television broadcast station.

Class II Cable Television Channel A signaling path provided by a cable television system to deliver to subscriber terminals television signals that are intended for reception by a television broadcast receiver without the use of an auxiliary decoding device and which signals are not involved in a broadcast transmission path.

Class III Cable Television Channel A signaling path provided by a cable television system to deliver to subscriber terminals signals that are intended for reception by equipment other than a television broadcast receiver or by a television receiver only when used with auxiliary decoding equipment.

Class IV Cable Television Channel A signaling path provided by a cable television system to transfer signals of any type from a subscriber terminal to another point in the cable television system.

Classified Ad Channel System (CACS) System by which still photographs and advertising information are transmitted via CATV to simulate newspaper classified advertisements.

Clear Channel Channel on which the dominant broadcasting stations render service over wide areas and which are cleared of objectionable interference within their primary service areas and over all or a substantial portion of their secondary service areas. Usually refers to AM broadcast stations.

CLI See "Cumulative Leakage Index."

Clicks (1) Short, sharp, undesired noises varying from light to heavy; (2) Impulse noise.

Climbing Space The vertical unobstructed 30-inch-square space reserved along the faces of a pole to provide access for linemen to place equipment and conductors on the pole.

Clipping The removal of that portion of a signal above or below a pre-set level.

Clock In a digital computer or control system, the device or control circuit which supplies the timing pulses that pace the operation of the digital system.

Closed-Circuit Television (CCTV) A private, usually in-plant television system not involving broadcasting but which transmits to one or more receivers through a cable.

Closed Loop System An electronic feedback control system wherein any residual error after correction is fed back directly into the system for inverse proportional control or correction to a normal condition.

Cluster Controller A device that can

control the input-output operations of more than one device connected to it. A cluster controller may be controlled by a program stored and executed within the unit.

Coax See "Coaxial Cable."

Coaxial Cable A type of cable used for broadband data and cable systems. Composed of a center conductor, insulating dielectric, conductive shield, and optional protective covering, this type of cable has excellent broadband frequency characteristics, noise immunity and physical durability. Synonymous with coax.

COBOL (Common Business-Oriented Language) An internationally accepted computer programming language.

Co-Channel Interference Interference on a channel caused by another signal operating on the same channel.

Code Computer language or program instructions.

Coding (of characters) (1) The bit combinations used to represent each character or symbol; (2) The process of writing computer language or program instructions. See also "Programming."

Color Banding A hue change usually from top to bottom of the bands. See also "Color Phase Shift Banding."

Color Burst In NTSC terminology, refers to a burst of approximately nine cycles of 3.58 MHz subcarrier on the back porch of the composite video signal. This serves as a color synchronizing signal to establish a frequency and phase reference for the chrominance signal.

Color Crosstalk Interference in a television picture caused by undesired mixing of chrominance and luminance information.

Color Flicker That flicker which results from fluctuation of both chrominance and luminance.

Color Fringing Spurious chromaticity at boundaries of objects in the picture.

Color Phase Shift Banding Banding made visible by differences in color phase between video head channels in a video tape recorder/player.

Color Picture Signal The electrical signal which represents complete color picture information, excluding all synchronizing signals.

Color Signal Any signal at any point in a color television system for wholly or partially controlling the chromaticity values of a color television picture.

Color Subcarrier In NTSC color, the 3.58 MHz subcarrier whose modulation sidebands are interleaved with the video luminance signal to convey color information.

Color Temperature The temperature at which a black body radiator must be operated to have a chromaticity equal to that of a light source.

Color Transmission A method of transmitting color television signals which can reproduce the different values of hue, saturation, and luminance which together make up a color picture.

Comb Generator A radio frequency (RF) signal generator that produces a series of output signals whose frequencies are mathematically related. This series, or comb, of frequencies is used as a phaselock reference signal for HRC or IRC headend processors and modulators.

Combiner See "Combining Network."

Combining Audio Harmonics The arithmetical sum of the amplitudes of all the separate audio harmonic components, which are multiples of the fundamental signal frequencies.

Combining Network A passive network which permits the combining of several signals into one output with a high degree of isolation between individual inputs; commonly used in CATV headends to combine the outputs of all processors and modulators into a single coaxial cable input. Synonymous with combiner.

Comité Consultatif International de Telegrafique et Telephonique (CCITT) See "International Telegraph and Telephone Consultative Committee."

Comité Consultatif International des Radio Communications (CCIR) See "International Consultative Committee for Radio."

Commercial Announcement (CA) Any advertising message for which a charge is made, or other consideration is received.

Commercial Continuity The advertising message of a program sponsor.

Common Carrier A telecommunications company, regulated by an appropriate government agency, that offers communications services to the general public via shared circuits at published tariff rates. In the United States, common carriers are regulated by the FCC or various state public utility commissions.

Communications Act of 1934 This act, passed by Congress in 1934, established a national telecommunications goal of high quality, universally available telephone service at reasonable cost. The act also established the Federal Communications Commission (FCC) and transferred federal regulation of all interstate and foreign wire and radio communications to this commission. It requires that prices and regulations for service be just, reasonable and not unduly discriminatory.

Communications Company Any organization legally authorized to provide transmission facilities from one point to another for lease or private communication purpose.

Communications Computer A computer that acts as the interface between another computer or terminal and a network, or which controls data flow in a network.

Communications Lines The conductors and their supporting or containing structures which are located outside of buildings and are used for public or private signal or communications service, and which operate at not exceeding 400 volts to ground or 750 volts between any two points of the circuit, and the transmitted power of which does not exceed 150 watts.

Communications Network The physical means for a group of nodes or hosts to interconnect or transmit data.

Communications Satellite An electronic retransmission vehicle located in space in a fixed earth orbit. Used by the cable television industry for transmission of its network programming, and by telephone companies for long distance voice and data traffic. Synonymous with bird.

Communications Satellite Corporation (COMSAT) A common-carrier service that provides commercial communications services.

Communications Satellite Corporation (COMSAT) A common-carrier service that provides commercial communications services.

Community Antenna Relay Service (CARS) (1) Designated microwave frequencies; (2) A microwave service band, operating at approximately 12 GHZ, for exclusive use of CATV.

Community Antenna Relay Station (CARS) A fixed station used for the transmission of television signals and related audio signals, and of standard and FM broadcast stations, from the point of reception to a terminal point from which the signals are distributed to the public by cable.

Community Antenna Television Association (CATA) Trade association for CATV operators and owners. Works in an advocacy position with policymakers and legislators at the state, regional, and national level. Generally represents independently owned and smaller multiple system operators.

Community Antenna Television System (CATV) See "Cable Television System."

Compacting Auger A mechanically powered screw auger used to bore tunnels under streets, sidewalks, driveways, or other obstacles.

Compander Combined word for compressor and expander in the transmission of audio frequency signals. Companding involves a volume compressor at the transmitter, and a volume expander at the receiver. By compressing or reducing dynamic range before transmission, and expanding or restoring dynamic range after reception, the noise immunity of the signal being transmitted is improved.

Compatibility Ability of one device to interconnect with another. In data communications, compatibility requires devices to have the same code, speed, and signal level.

Compile (1)To translate a computer program expressed in a problem-oriented language into a computer-oriented language; (2) To prepare a machine language program from a computer program written in another programming language by making use of the overall logic structure of the program, or by generating more than one computer instruction for each symbolic statement, or both, as well as performing the function of an assembler.

Compiler A systems program that translates high-level language programs into machine-level code.

Complementary Metal-Oxide Semiconductor (C-MOS) Chips that use far less electricity than other types where circuits are relatively immune to electrical interference and operate in a wide range of temperatures. C-MOS transistors on the chip are paired, with one requiring positive voltage and the other negative voltage to work. The transistors thus offset, or complement, each other's power requirements.

Composite Color Signal The color picture signal plus blanking and all synchronizing signals.

Composite Second Order Beat (1) A clustering of second order beats 1.25 MHz above the visual carriers in cable systems; (2) A ratio, expressed in decibels, of the peak level of the visual carrier to the peak of the average level of the cluster of second-order distortion products located 1.25 MHz above the visual carrier.

Composite Triple Beat (1) A clustering of third-order distortion products around the visual carriers in cable systems; (2) A ratio, expressed in decibels, of the peak level of the visual carrier to the peak of the average level of the cluster of third-order distortion products centered around the visual carrier.

Composite Video Signal The complete video signal. For monochrome, it consists of the picture signal and the blanking and synchronizing signals. For color, additional color synchronizing and color picture information are added.

Compression (1) A less-than-proportional change in output for a change in input; (2) The reduction in amplitude of one portion of a wave-form relative to another portion.

Compressor A circuit or device which limits the amplitude of its output signal to a predetermined value in spite of wide variations in input signal amplitude. It effectively reduces the dynamic range of the original input signal.

Compulsory License A license granted to cable television systems for the retransmission of television and radio broadcast signals that is conditioned upon compliance with Federal Communications Commission regulations and the remittance of royalty payments to the U.S. Copyright Office. The royalty fee, which is later distributed to the copyright owners of programs carried on the signals, is higher for larger cable systems and is based on a sliding scale percentage of fees received for television and radio broadcast services provided to subscribers.

CompuServe Information Services (CIS) A computer network service.

Computer A functional unit that can perform substantial computations, including numerous arithmetic operations or logic operations, often without intervention by a human operator. See also "Central Processing Unit."

Computer-Aided Design (CAD) A computer system whereby engineers create a design and see the proposed product in front of them on a graphics screen or in the form of a computer printout.

Computer-Aided Engineering Essentially computer software purporting to use the computer to predict how the part, machine or manufacturing process can perform.

Computer-Assisted Instruction (CAI) A data-processing application in which a computing system is used to assist in the instruction of students. The application usually involves a dialog between the student and a computer program which informs the student of mistakes as they occur.

Computer-Forced Tune A condition where the channel selected by the terminal is dictated by the computer center and not by the subscriber.

Concentrator In communications systems, a functional unit that permits a common path to handle more data sources than there are channels currently available within the path.

Conduit A tube, manufactured of a protective material, through which CATV or other cable is conveyed in an underground system.

Cone See "Safety Cone."

Connect Time Time period during which a user is utilizing a computer on-line.

Contrast The range from white to black in a scene or television picture.

Control Circuit A telephone, telegraph, or radio circuit used to provide a direct link to coordinate activities at or between the program source and control points.

Converter Device for changing the frequency of a television signal. A cable headend converter changes signals from frequencies at which they are broadcast to clear channels which are available on the cable distribution system. A set-top converter is added in front of a subscriber's television receiver to change the frequency of the midband, superband, or hyperband signals to a suitable channel or channels (typically a low VHF channel) which the television receiver is able to tune.

Converter Disable A state where a normal TV signal output is not available from the terminal. The method of reaching this state may vary upon the condition of the terminal.

Co-Op Fees paid in part by a program supplier and in part by the cable operator for a CATV sales promotion.

Copyright Royalty Payments made by CATV operators to the Copyright Royalty Tribunal for distribution to original copyright holders for certain television programming, in lieu of direct payments per program.

Copyright Royalty Tribunal Organization responsible for collecting copyright payments from operators and distributing them to holders of copyright for programs that appear on CATV systems.

Core Memory A nearly obsolete type of central processing unit memory that stores information on magnetically charged, doughnut-shaped cores made of ferrite and lithium. Core memories have largely been superseded by semiconductor memories.

Core Storage A magnetic storage in which the magnetic medium consists of magnetic cores.

Corner Reflector Antenna An antenna which employs a piece of folded metal or mesh reflector or group of rods mounted to resemble a folded piece of metal, to increase the antenna gain in the unobstructed direction.

Courseware Software used in teaching. Often used to describe computer programs designed for the classroom.

CPR (1) Cardiopulmonary resuscitation; (2) Capital purchase request.

CPS (1) Characters per second; (2) Cycles per second (Hz).

CPU See "Central Processing Unit."

Crash An abrupt, unplanned computer system shutdown caused by a hardware or software malfunction.

Crawl Space (1) Space for textual messages usually at the bottom of the television screen; (2) Area under a house, mobile home, or other building, usually for access to utilities, plumbing, and heating/cooling.

Critical Distance The length of a particular cable which causes worst case reflection if mismatched; depends on velocity of propagation and attenuation of cable at different frequencies.

CRO See "Cathode-Ray Oscilloscope."

Cropping The elimination of picture information near the edge or edges of a picture.

Cross Assembler A program used with one computer to translate instructions for another computer.

Cross Modulation A form of television signal distortion where modulation from one or more television channels is imposed on another channel or channels.

Cross-Compiler A compiler that runs on a computer other than the one for which it was designed to compile code.

Cross-Ownership The ownership of two or more kinds of communications services in the same market, e.g., newspaper and television stations, by a single entity.

Crosstalk (1) Undesired transfer of signals from one circuit to another circuit; (2) The phenomenon whereby a signal transmitted on one circuit or channel of a communications system is detectable or creates an undesirable effect in another circuit or channel.

CRT See "Cathode-Ray Tube."

Cryogenic Storage A storage device that uses the superconductive and magnetic properties of certain materials at very low temperatures.

Cryogenics The study and use of devices utilizing properties of materials near absolute zero in temperature.

CSR See "Customer Service Representative."

CTAM See "Cable Television Administration and Marketing Society."

Cuckoo Wavetek's trade name for a specific type of signal leakage detection equipment. The stepped sound of the transmitted tone is similar to a cuckoo clock.

Cue Audio A second audio channel which may be recorded independently and is usually used for recording direction cues, explanatory notes or in some cases, for control signals.

Cue Circuit A one-way communication circuit used to convey program control information.

Cumulative Leakage Index (CLI) A figure of merit derived mathematically from the number and severity of signal leaks in a cable system. Compliance with FCC regulations requires a CLI figure of merit of 64 or less.

Cursor A symbol on the display of an editing or display terminal that can be moved up, down, or sideways and indicates where the next character is to be located, or where "home" or beginning is located.

Customer Service Providing telephone and in-home assistance for CATV system customers.

Customer Service Representative (CSR) An individual employed by the cable company to answer the telephone, write service and installation orders, answer customers' questions, receive and process payments, and perform other customer service-related activities.

Cut (1) An undesired interruption in the transmission of program material. Loss of audio and video signals; (2) Command to immediately stop transmission or recording of audio and/or video material.

Cut-Off Frequency That frequency beyond which no appreciable energy is transmitted.

Cut Start (Up-Cut) The commencement of transmission of the video and audio signals of a program which is already in progress.

Cut to Time (Down-Cut) The termination of a program before its completion in order to comply with the time period schedule for that program.

Cycle One complete alternation of a sound or radio wave. The rate of repetition of cycles is the frequency. See also "Hertz."

D

Daisy Print Wheel A plastic or metal print wheel found in word-processing printers that makes the typing impression on paper. Its unique circular design allows these units to print up to 540 words per minute.

Damped Oscillation Oscillation in which the amplitude of each peak is lower than that of the preceding one; the oscillation eventually decays to zero.

Damping A characteristic built into electrical circuits and mechanical systems to prevent unwanted oscillatory conditions.

Data (1) A representation of facts, concepts, or instructions in a formalized manner suitable for communication, interpretation, or processing by human or automatic means; (2) Information.

Data Acquisition The process of identifying, isolating, and gathering source data to be centrally processed.

Data Bank A comprehensive collection of libraries of data. For example, one line of an invoice may form an item, a complete set of such records may form a file, the collection of inventory control files may form a library, and the libraries used by an organization are known as its data bank.

Data Communications (1) The movement of encoded information by means of electrical or electronic transmission systems; (2) The transmission of data from one point to another over communications channels.

Data Compression A technique that saves storage space by eliminating gaps, empty fields, redundancies, or unnecessary data to shorten the length of records or blocks.

Data Network Telecommunications network built specifically for data transmission, rather than voice transmission.

Data Phone A unit that permits data to be transferred over a telephone line.

Data Processing The systematic performance of operations upon data; for example, handling, merging, sorting, computing.

Data Switching Exchange Telecommunications switching station built specifically for data network transmission control.

Database A collection of information in a form that can be manipulated by a computer and retrieved by a user through a terminal.

dB See "Decibel."

dBc Decibel-carrier. A ratio expressed in decibels that refers to the gain or loss relative to a reference carrier level.

dBd Decibel-dipole. A ratio expressed in decibels that refers to the gain or loss relative to a dipole antenna.

dBi Decibel-isotropic. A ratio expressed in decibels that refers to the gain or loss relative to an isotropic antenna.

dBm See "Decibel Milliwatt."

dBmV See "Decibel Millivolt."

DBS See "Direct Broadcast Satellite."

dBV See "Decibel Volt."

dBW See "Decibel Watt."

Dead Time Any delay deliberately placed between two related actions in order to avoid overlap that can confuse or permit a particular different event, such as a control decision, switching event, or similar action, to take place.

Dead Zone The range of input values for a signal that can be altered but has no impact on the output signal.

Debug To detect, trace, and eliminate mistakes in computer programs or in other software.

Decibel (dB) A unit that expresses the ratio of two power levels on a logarithmic scale.

Decibel Millivolt (dBmV) A unit of measurement referenced to one millivolt across a specified impedance (75 ohms in CATV).

Decibel Milliwatt (dBm) A unit of measurement referenced to one milliwatt across a specified impedance.

Decibel Volt (dBV) A unit of measurement referenced to one volt across a specified impedance.

Decibel Watt (dBW) A unit of measurement referenced to one watt across a specified impedance.

Decoder Electronic device which translates scrambled or decoded signals in such a way as to recover the original message or signal. Synonymous with descrambler and decryptor.

Decryptor See "Decoder."

Dedicated Machines, programs, or procedures designed or set apart for special or continued use.

Dedicated Channel (1) Telephones: A channel that is not switched; (2) CATV: A channel reserved for future use.

Dedicated Circuit A circuit designated for exclusive use by two users.

Dedicated Connection Out-of-use term for non-switched connection.

Dedicated Device A device that cannot be shared among users.

Dedicated Line See "Leased Line."

Dedicated Port The access point to a communication channel used only for one specific type of traffic.

Dedicated Service A communication link reserved exclusively for one user.

De-emphasis Required departure from a flat gain/frequency characteristic in part of a facility because of the use of pre-emphasis earlier in the facility.

Default A value, attribute, or option that is assumed when no alternative has been specified.

Definition Distinctness or clarity of picture. See also "Resolution."

Degausser (1) Demagnetizer; (2) A device for bulk erasing magnetic tape.

Delay Counter A counter for inserting a deliberate time delay allowing an operation external to the program to occur.

Delay Distortion Distortion resulting from non-uniform velocity of transmission of the various frequency components of a signal through a transmission system.

Delay Line A line or network designed to introduce a desired delay in the transmission of a signal, usually without appreciable distortion.

Delivery Time The time interval between the beginning of transmission at an initiating terminal and the completion of reception at a receiving terminal.

Demodulate To retrieve an information-carrying signal from a modulated carrier. See also "Modem" and "Modulate."

Demodulator A device that removes the modulation from a carrier signal.

Demographics Statistical breakdown by categories (e.g., age, sex) of a particular section of the viewing public.

Depth of Field The range of distance in which things appear in focus to a camera.

Descrambler See "Decoder."

Designated Community In a major television market, a community, listed in FCC regulations, commonly referred to as a "top 100" market community.

Detail The most minute elements in a picture which are distinct and recognizable. Similar to definition or resolution.

Diagnostic A computer program for automatically debugging or assisting other programs or for finding the cause of hardware failures. Synonymous with diagnostic program and diagnostic routine.

Diagnostic Program See "Diagnostic."

Diagnostic Routine See "Diagnostic."

Didon An Antiope-based broadcast teletext system used in France.

Dielectric A non-conductive insulator material between the center conductor and shield of coaxial cable. The dielectric constant determines the propagation velocity.

Differential Analyzer An analog computer using interconnected integrators to solve differential equations.

Differential Delay The difference in the delays experienced by two sinusoids of different frequencies passing through a communications channel.

Differential Gain The difference in gain of a video facility at a subcarrier frequency between any two luminance levels from blanking to reference white level.

Differential Phase The maximum difference in phase of a video facility at the color subcarrier frequency between any two luminance levels from blanking to reference white level.

Diffraction Region The region lying adjacent to and below the radio transmission horizon.

Diffused Illumination Diffused light which illuminates a relatively large area with an indistinct beam.

Digital Descriptive of any process that uses discrete levels (usually 0 and 1) to represent characters or numbers.

Digital Computer A computer that operates on discrete data by performing arithmetic and logic processes on these data.

Diode An electronic device used to permit current flow in one direction and to inhibit current flow in the other.

Diplexer See "Diplexing Filter."

Diplexing Filter A device that provides

signal branching on a frequency division basis. Synonymous with diplexer.

Dipole Antenna A straight, center-fed one-half wavelength antenna.

Direct Access The ability to obtain data from a storage device, or to enter data into a storage device, in such a way that the process depends only on the location of that data and not on a reference to data previously accessed.

Direct Broadcast Satellite (DBS) A satellite service of one or more entertainment or information program channels which can be received directly using an antenna on the subscriber's premises.

Direct Coupling A means of connecting electronic circuits or components so that the amplitude of currents within each are independent of the frequency of those currents.

Direct Pickup Unwanted signal ingress usually from over-the-air television broadcast stations, translators or FM radio stations directly into the cable system, or the subscriber's television set or FM receiver.

Direct Read After Write (DRAW) A laser-based technology for recording data on a videodisk.

Directional Coupler A passive signal splitting device, with minimum signal loss between the input port and the output port (through loss), a specified coupling loss between the input port and the tap port (tap or coupler loss), and very high loss between the output port and tap port (isolation).

Directional Illumination Directional light which illuminates a very small area with a light beam.

Directional Tap See "Multitap."

Discrete Pertaining to data in the form of distinct elements such as characters, or to physical quantities having distinctly recognizable values.

Discrete Component A self-contained

device which offers one particular electrical property or function in lumped form, that is, concentrated in one place in a circuit; it exists independently, not in combination with other components (e.g., transistor, resistor, capacitor).

Dish (Antenna) A transmitting or receiving antenna shaped like a dish; used to receive radio and television signals from a communications satellite or microwave link.

Disk Drive A computer data storage device in which data is stored on the magnetic coating (similar to that on magnetic tape) of a rotating disk.

Disk Storage See "Disk Drive."

Diskette A flexible magnetic disk used in recording, as in computers and data storage systems.

Display The visual presentation on the indicating device of an instrument.

Distant Signal The signal of a television broadcast station which is extended or received beyond the Grade B Contour of that station.

Distortion An undesired change in waveform of a signal in the course of its passage through a transmission system.

Distributed Data Processing Data processing in which some or all of the processing, storage, and control functions, in addition to input-output functions, are situated in different places and connected by transmission facilities.

Distributed Function The use of programmable terminals, controllers, and other devices to perform operations that were previously done by the processing unit, such as managing data links, controlling devices, and formatting data.

Distribution Amplifier See "Bridging Amplifier."

Distribution Processing (1) Computer processing systems in which the control functions and/or computing functions are shared among several network nodes; (2) A single logical set of processing functions implemented across a number of computers. A central facility may or may not be part of the network.

Distribution System The part of a CATV system consisting of trunk and feeder cables which are used to carry signals from the system headend to subscriber terminals. Often applied, more narrowly, to the part of a CATV system starting at the bridger amplifiers. Synonymous with trunk and feeder system.

Distribution Tap-Off A passive device used to connect subtrunks or feeder cables to the main trunk.

Diversity Reception A method of preventing or minimizing the effects of signal fade by using receivers whose antennas are five to ten wavelengths apart. Each receiver, tuned to the same signal, feeds a common audio amplifier, which reduces the likelihood of overall fading since the extent of signal fade is different at the various antennas.

Door-to-Door Traditional method of selling CATV services by sending representatives of the cable company to each home in a franchise area.

Down Inoperable; not functioning.

Down in the Noise A signal that tends to be very weak, or whose strength is about equal to interfering signal strengths.

Down Time The period during which electronic equipment is completely inoperable.

Downconverter A type of radio frequency converter characterized by the frequency of the output signal being lower than the frequency of the input signal. Synonymous with input converter.

Downgrade The discontinuance, by a subscriber, of a premium program service or any other added service or product from the existing level of CATV service.

Downlink Transmission of signals from a satellite to a dish or earth station.

Download Transfer data from a main computer or memory to a remote computer or terminal.

Downstream In a cable system, the direction of signal transmission from the headend to subscriber terminals.

DRAW See "Direct Read After Write."

Drift A change in the output of a circuit that occurs slowly.

Drip Loop An intentional loop formed in the subscriber drop cable to prevent precipitation from following the cable and entering equipment or the subscriber's home.

Drop The line from the feeder cable to the subscriber's television or converter.

Drop Cable (1) Cable that connects a vertical riser to the modems interfacing with user's end equipment; (2) The coaxial cable that connects the feeder portion of the distribution system to the subscriber's premises.

Drop-Outs Black or white lines or spots appearing in a television picture originating from the playback of a video tape recording.

Dual Cable System See "Multiple Cable System."

Duplex In a communications channel, the ability to transmit in both directions. See also "Half Duplex," "Full Duplex," and "Simplex."

Duty Cycle The operation of machines or devices; denotes the ratio of "on-time" to the total of one operating cycle.

Dynamic Gain A change in transmission system gain as measured by changes in peak-to-peak luminance and sync levels resulting from variations of the average picture level.

Dynamic Range The ratio (in decibels) of the weakest or faintest signals to the strongest or loudest signals reproduced without significant noise or distortion.

E

E-Layer A heavily ionized signal reflecting region located 50-70 miles above the surface of the earth, within the ionosphere.

Early Finish The completion of program material before the end of the period designated for that program.

Early Start The commencement of the transmission of program material before the starting time scheduled for that program period.

EAROM See "Electrically Alterable Read Only Memory."

Earth Station (Dish) A parabolic antenna and associated electronics for receiving or transmitting satellite signals.

Easement CATV:the right to use the land of another for a specific purpose such as to pass over the land with cables; the right of ingress and egress over the land of another.

Echo A signal that has been reflected or otherwise returned.

Edge Effect The overemphasizing of well-defined objects due to the addition of leading black or leading white outlines to the objects.

Editor A computer program used to edit (prepare for processing) text or data.

Editorial Program Programs presented for the purpose of stating opinions of the licensee.

Educational Access Channel A cable television channel specifically designated for use by local education authorities.

Educational Broadcasting Corporation Owner and licensee of New York City's public television station, Channel 13, WNET. Produces programs that are distributed by the Public Broadcasting Service (PBS) to more than 250 noncommercial television stations throughout the United States.

Educational Program Includes any program prepared by, on behalf of, or in cooperation with, educational institutions, libraries, museums, PTAs or similar organizations.

Effective Field The root-mean-square (RMS) value of the inverse-distance intensity field strength voltage at a distance, usually one mile, from the broadcast antenna in all directions in the horizontal plane.

Effective Height The above-ground height of an antenna in terms of its performance as a transmitting or receiving device (center of radiation) as opposed to its physical height above ground.

Effective Radiated Power (ERP) The product of the antenna power input times the antenna power gain or the antenna field gain squared. Where circular or elliptical polarization is employed, the term is applied separately to the horizontal and vertical components of radiation.

EFTS See "Electronic Funds Transfer System."

Egress In CATV, unwanted leakage of signals from a cable system. See also "Leakage."

EIA See "Electronic Industries Association."

Electrical Length The effective length of an antenna, transmission line, or device in terms of actual performance, expressed in wavelengths, radians, or degrees. The electrical length is usually different than the actual length due to ground-capacitance effects, end effects and the velocity of propagation of electromagnetic waves in wire.

Electrically Alterable Read Only Memory (EAROM) A type of memory that is nonvolatile, like ROM, but can be altered, or have data written into it, like RAM.

Electromagnetic Interference Any electromagnetic energy, natural or manmade, which may adversely affect performance of the system.

Electromagnetic Spectrum The frequency range of electromagnetic radiation that includes radio waves, light and X-rays. At the low frequency end are subaudible frequencies (e.g., 10 Hz) and at the other end, extremely high frequencies (e.g., X-rays, cosmic rays).

Electronic Editing The process by which audio and/or video material is added to a previously recorded tape in such a manner that continuous audio and/or video signals result.

Electronic Funds Transfer System (EFTS) A generic term for banking communications systems.

Electronic Industries Association (EIA) A U.S. association providing standards, including interfaces, designed for use between manufacturers and purchasers of electronic products.

Electronic Mail A message service which uses computers and telecommunications links to deliver information. Synonymous with E-mail.

Emergency Override An emergency communications system which allows messages or announcements to replace the normal picture and/or sound on all channels on a cable system.

Emergency Power Generator or battery back-up power to replace primary power during an electrical outage.

Emitter Tuner Principle A circuit which extends the high frequency response of transistors by using variable trimmer capacitors to neutralize detrimental emitter inductance.

Emulator A program that allows one processor to simulate the instruction set of another processor.

Encoder See "Scrambler."

Encryption Coding of a signal for privacy protection, in particular when transmitted over telecommunication links.

End-Fire Array An antenna of multiple elements where the principle direction of signal radiation coincides with the direction of the antenna array axis or plane.

End of Tape Sensing A form of sensing (optically or mechanically) that automatically stops the tape transport at the end of tape or upon breakage of tape.

End User The intended final user of a circuit, device or information.

ENG Electronic news gathering.

Engineering Service Circuit (ESC) A voice or telegraph circuit interconnecting stations in a network. For the use of operations and maintenance personnel of the commmunications companies when coordinating and maintaining services on the network.

Enter To place on the line a message to be transmitted from a terminal to the computer.

Entertainment and Sports Programming Network (ESPN) A satellite-delivered sports network.

Entertainment Program Includes all programs intended primarily as entertainment.

EPROM Erasable-Programmable Read-Only Memory. See "Erasable Read-Only Memory."

Equalization Adjusting the frequency response of an amplifier or network so that it will affect all signal components within a specific bandwidth to result in a desired overall frequency response.

Equalizer A passive device or circuit with a tilted frequency response opposite that of the cable preceding it, to compensate for the response of that cable.

Equalizing Pulses of one-half the width of the horizontal sync pulses which are transmitted at twice the rate of the horizontal sync pulses during the blanking intervals immediately preceding and following the vertical sync pulses. The action of these pulses causes the vertical deflection to start at the same time in each interval, and also serves to keep the horizontal sweep circuits in step during the vertical blanking intervals immediately preceding and following the vertical sync pulse.

Erasable Read-Only Memory (EROM) In a computer, the read-only memory

(ROM) that can be erased and reprogrammed. Synonymous with Erasable-Programmable Read-Only Memory (EPROM).

Erasable-Programmable Read-Only Memory (EPROM) See "Erasable Read-Only Memory."

EROM See "Erasable Read-Only Memory."

ERP See "Effective Radiated Power."

ESC (Order Wire) See "Engineering Service Circuit."

ESPN See "Entertainment and Sports Programming Network."

ETV Educational television.

European Conference of Postal and Telecommunications Administrations (CEPT) (Conference Europeene des Administrations des Postes et de Telecommunications) Organization of postal and telecommunications administrations of the western European countries. Stresses information and expertise exchange among members, and the simplification, improvement, and coordination of postal and telecommunications services.

Eurovision An arrangement between European and Mediterranean television services for the distribution of television programs.

Exclusive Franchise A franchise that allows the construction and operation of only one cable television system within the bounds of its governmental authority.

Exclusivity A provision in a film or program contract that grants exclusive playback rights for that film or program to a television station in a given market.

Execute To perform the operations required by an instruction, command or program.

Expansion (1) Increase in amplitude of a

portion of the composite video signal relative to that of another portion; (2) In companding, the restoration of compressed audio signals to their original dynamic range.

Expansion Loop A loop intentionally formed in coaxial cable to compensate for temperature-caused expansion and contraction of the cable.

Experimental Period That time between 12:00 midnight and local sunrise for broadcast station transmission.

Eye Anchor rod eye. The attachment point of an anchor rod.

F

"F"-Type Connectors A connector used by the cable television industry to connect coaxial cable to equipment.

FAC See "Facsimile."

Facility The site, including land and buildings containing all or part of a system or systems of technical apparatus; for example, a headend facility or microwave relay facility, employed for purposes of electronic communication or data processing.

Facsimile (FAC) (FAX) A system for the transmission of images.

Facsimile Index of Cooperation The product of the number of lines per inch, the available line length in inches and the reciprocal of the line-use ratio.

Facsimile Line-Use Ratio The ratio of the available line to the total length of scanning line.

Facsimile Rectilinear Scanning The process of scanning an area in a predetermined sequence of narrow, straight parallel strips.

Fading A fast or slow deterioration of signal quality due to increasing loss in an electromagnetic propagation path.

Fairness Doctrine FCC requirement that any CATV system engaging in cablecasting shall afford reasonable opportunity for the discussion of conflicting views on issues of public importance.

Fast Lap Switching Video switching by use of a solid-state switching element.

Fault An accidental condition that results in a functional unit failing to perform in a required manner.

FAX See "Facsimile."

FCC See "Federal Communications Commission."

FD See "Floppy Disk."

Federal Communications Commission (FCC) The U.S. government agency established in 1934 to regulate electronic communications. The FCC succeeded the Federal Radio Commission.

Feeder Cables The coaxial cables that take signals from the trunk line to the subscriber area and to which subscriber taps are attached. Synonymous with feeder line.

Feeder Line See "Feeder Cables."

Feedforward An amplification technique which provides improved distortion performance and output capability compared to conventional push-pull amplification techniques.

Fiber Optics The technology of guiding and projecting light for use as a communications medium. Hair-thin glass fibers which allow light beams to be bent and reflected with low levels of loss and interferences are known as "glass optical wave guides" or simply "optical fibers."

Field One-half of a complete picture (or frame) interval, containing all of the odd or even scanning lines of the picture.

Field Angle The angle of the light beam which covers an area at whose edges the candle power is 10 percent of the maximum candle power.

Field Blanking Interval The period provided at the end of field picture signals primarily to allow time for the vertical sweep circuits in receivers to return the electron beam completely to the top of the raster before the picture information of the next field begins.

Field Frequency The rate at which a complete field is scanned, nominally 60 times per second for NTSC monochrome video signals, and 59.94 times per second for NTSC color video signals.

Field Intensity (1) The strength of an electric or magnetic field; (2) The strength of a radio wave, usually expressed in microvolts, millivolts, volts, microvolts per meter, millivolts per meter, or volts per meter.

Field Side The side of the utility pole opposite the vehicle access side.

Field Strength The intensity of an electromagnetic field at a given point, usually referred to in microvolts per meter.

Field Strength Meter (FSM) A frequency selective heterodyne receiver capable of tuning to the frequency band of interest; in cable television, 5 to 550 MHz with indicating meter showing the magnitude input of voltage and a dial indicating the approximate frequency. Synonymous with signal level meter.

Field Time Distortion Linear wave-form distortion occurring in the time domain of the television field.

FIFO See "First In-First Out."

Figure 8 Cable Coaxial cable manufactured with an integrated messenger cable.

Figure of Merit (1) Relative term indicating an amplifier's comparative performance characteristics; (2) With regard to the cumulative leakage index (CLI) in a cable system, figure of merit relates to the index that indicates the overall signal leakage performance of the cable plant.

File An organized collection (in or out of sequence) of records related by a common format, data source or application.

Fill Light Generally diffused light to reduce shadow or contrast range. It can be directional.

Film Break Interruption of the transmission of program material due to the mechanical failure of film being used as a source of material.

Film Chain Equipment to transfer film movies or slides to video.

Filter A passive circuit or device which passes one frequency or frequency band while blocking others, or vice versa.

Firmware Software stored in read-only memory (ROM). Synonymous with microcode.

First In-First Out (FIFO) Processing arrangement used in data manipulation wherein the first data input is the first output.

First Line Hue Shift Banding Banding made visible by a difference in hue of the first line of each band.

First Run Non-Series Programs Programs, other than series, that have no national network television exhibition in the United States and no regional network exhibition in the relevant market.

Fixed Field A computer record subdivision which contains an allocated number of specific characters or information units.

Fixed Receiver A satellite receiver, usually crystal controlled, that can receive only one transponder from a communications satellite.

Fixed Storage A storage device whose contents are inherently non-erasable, non-erasable by a particular user, or non-erasable when operating under particular conditions; for example, a storage controlled by a lockout feature, or a photographic disk.

Flag (1) A piece of information added to a data item that provides information about the data item; (2) A character or code that signals the presence of some condition.

Flash Momentary disturbance of a major area of a television picture of such duration that the real impairment cannot be readily identified.

Flat Loss Equal signal loss, or attenuation, at all frequencies.

Flat Output Operation of an amplifier where all output signals are at the same amplitude.

Float Shifting characters into position to the right or left as determined by data structure or programming devices.

Float-Point Register A register used to manipulate data in a floating-point representation.

Flooded Cable Coaxial cable that has a coating of flooding compound between the shield and jacket.

Flooding Compound A viscous, non-hardening, non-drying material that is placed between the shield and jacket of coaxial cable to provide water proofing and sealing properties.

Floppy Disk (FD) Out-of-use term for diskette.

Floppy Disk Controller Device that provides control of data transfer to and from a diskette.

Fluorescence Emission of light from a substance during, or as a result of, excitation.

Flutter (1) Rapid fluctuations in the pitch of a reproduced sound; (2) Rapid fluctuations in received signal strength.

FM Broadcast Band The band of frequencies extending from 88 to 108 MHz.

FM Broadcast Station A radio station employing frequency modulation in the FM broadcast band and licensed primarily for the transmission of sound emissions intended to be received by the general public.

FM Modulator In CATV, a device similar to an FM transmitter that is used to cablecast signals in the FM band on a cable system.

FM Stereophonic Broadcast The transmission of a stereophonic program by a single FM broadcast station, utilizing the main channel and a stereophonic subchannel, to provide two separate sound channels.

Focus Maximum convergence of the electron beam manifested by minimum spot size on the phosphor screen; registering picture elements in sharp definition.

Focus Group A group of consumers who have been brought together to discuss a topic or product. Using focus groups is a well-known method of doing qualitative market research.

Following (or Trailing) Blacks A picture condition in which the edge following a white object is overshadowed toward black.

Following (or Trailing) Whites A picture condition in which the edge following a black or gray object is shaded toward white.

Footcandle The unit of illumination equal to 1 lumen per square foot.

Footprint The area of the earth's surface to which a satellite transmits.

Forward Direction The direction of the signal flow away from the headend. High frequencies are amplified in this direction. See also "Downstream."

Fourier Series A mathematical analysis permitting any complex wave-form to be resolved into a fundamental plus a finite number of terms involving its harmonics.

Frame One complete picture consisting of two fields of interlaced scanning lines.

Frame-Grabber The logic element of a broadcast teletext decoder that "captures" a designated, numbered frame as it is transmitted.

Frame Roll A momentary, vertical rolling of the picture due to instability in or loss of the vertical synchronization portion of the video signal.

Frame Store A video storage and display technique where a single frame of video is digitized and stored in memory for retrieval and subsequent display or processing.

Franchise Authorization issued by a municipal, county, or state government entity which allows the construction and operation of a cable television system within the bounds of its governmental authority.

Franchise Area The geographical area specified by a franchise where a cable operator is permitted to provide CATV service.

Franchise Fee Fee paid by a cable

operator to a government authority(ies) for the right to operate the cable franchise in a specified area.

Franchising Authority The municipal, county, or state government entity that grants a cable operator a franchise to construct and operate a cable television system within the bounds of that entity's governmental authority.

Free Running Frequency The frequency at which a synchronized generator (e.g., an oscillator) will operate when the synchronizing signal is removed.

Free Space Field Intensity The radio field intensity that would exist at a point in the absence of waves reflected from the earth or other reflecting objects. Synonymous with free space field strength.

Free Space Field Strength See "Free Space Field Intensity."

Free Space Region The zone along the propagation path that is free from objects that might absorb or reflect radio energy.

Frequency The number of complete alternations of a sound or radio wave in a second, measured in Hertz. One Hertz equals one cycle per second.

Frequency Agility The ability to easily tune to other frequencies.

Frequency Band Splitter/Mixer A device similar to other splitters except that it provides branching on a frequency division basis.

Frequency Diversity The use of more than one frequency to transmit identical information to overcome fading and interference and to improve signal transmission reliability.

Frequency Division Multiplexing See "Wavelength Multiplexing."

Frequency Modulation (FM) A form of modulation in which the frequency of the carrier is varied in accordance with the instantaneous value of the modulating signal.

Frequency Range That range of frequencies over which a device performs or meets its specifications.

Frequency Response The gain vs. frequency characteristic of a circuit, device or network.

Frequency Shift Keying (FSK) (1) A form of frequency modulation in which the modulating signal shifts the output signal between predetermined values. The instantaneous frequency is shifted between two discrete values, often called mark and space frequencies; (2) A signalling method in which different frequencies are used to represent different characters to be transmitted.

Frequency Translation Conversion from one channel or frequency to another.

Front Porch That portion of the composite picture signal which lies between the trailing edge of action video and the leading edge of the next corresponding sync pulse.

FSK See "Frequency Shift Keying."

FSK Failure A state when a normal downstream data reception of modulated FSK signal is absent.

FSM See "Field Strength Meter."

Full Duplex An electronic circuit or network that permits simultaneous transmission of signal in two directions.

Full Network Station A commercial television broadcast station that generally carries in weekly prime time hours 85 percent of the hours of programming offered by one of three major national television networks with which it has a primary affiliation.

Full Tilt The operation of an amplifier with a response that is tilted or sloped sufficiently to compensate for the response of the cable following that amplifier. This technique results in equal amplitudes on all signals at the input to the next amplifier.

Fully Integrated System A CATV system designed to take advantage of the optimum amplifier-cable relationship for highest performance at lowest cost, and including the density of subscriber taps required.

Funny Paper Effect See "Lagging Chrominance" and "Leading Chrominance."

Fused Disk Cable A type of coaxial cable that has an air dielectric and uses plastic disks to support the center conductor.

G

G-Line An obsolete technology using a single conductor wire for long distance television signal transmission.

Gain A measure of the signal level increase in an amplifier, usually expressed in decibels.

Gain/Frequency Distortion Distortion which results when all of the frequency components of a signal are not transmitted with the same gain or loss.

Gate A combinational circuit with only one output channel.

Gate Pulse Extended duration signals designed to increase the possibility of coincidence with other pulses. Gate pulses present with other pulses cause circuits or units to perform intended operations.

General Purpose Language A programming language that is not restricted to a single type of computer; for example, BASIC, FORTRAN.

Geostationary Describes a satellite in orbit 22,300 miles above the equator that revolves around the earth with an angular velocity equal to that of the earth's rotation about its own axis. The satellite's position relative to the earth's surface is constant (stationary), so little or no ground antenna tracking is needed. Synonymous with geosynchronous.

Geosynchronous See "Geostationary."

Ghosts Outlining or double images on a television picture, usually caused by signal path reflections.

GHz See "GigaHertz."

GigaHertz (GHz) One billion (10^9) cycles per second.

G-Line An obsolete technology using a single conductor wire for long distance television signal transmission.

Glitch (1) A narrow horizontal bar moving vertically through a television picture; (2) A short duration pulse moving through the video signal at approximately reference black level on a waveform monitor; (3) A random error in a computer program; (4) Any random, usually short, unexplained malfunction.

Golden Ace Award given annually for highest achievement in original, made-for-cable programs. See also "Award for Cablecasting Excellence."

Goodnight Time The term used to

designate the actual termination time of video or associated audio transmissions at the customer's facility.

Governmental Channel That channel set aside for free local government use during a system's developmental phase, a period of five years after subscriber service begins or the basic trunk line is completed. The allocating of free governmental channels is mandated by the Federal Communications Commission for systems operating in the top 100 markets.

Governmental Cablecasting The preparation and distribution of information over CATV/public access channels by government agencies.

Governmental Cablecasting Network Coordinated cable system operator network, set up to cycle government information to all participants in the network.

Grade A Service The zone or condition in which the quality of broadcast reception is expected to be satisfactory to an average observer at least 70 percent of the time for at least 90 percent of the receiving locations.

Grade B Service The zone or condition in which the quality of broadcast reception is expected to be satisfactory to an average observer at least 90 percent of the time for at least 50 percent of the receiving locations. Synonymous with predicted grade B contour.

Grain or Graininess (1) A uniform distribution of spots throughout the television picture from a motion picture source; (2) Noise in the picture.

Grandfather To exempt an entity from certain regulations because that entity existed prior to those regulations, or operated under rules or conditions that were in effect prior to those regulations.

Granted Date Official authorization date of station license or construction permit by the Federal Communications

Commission.

Graphic Display Unit A communications terminal that displays data on a screen in a graphic format.

Graphic Symbols Short form of "The Graphic Symbols for Cable Television Systems," the established standard for the schematic symbols used in cable system design and layout.

Gray Scale A test chart, slide or electronically produced wave-form, consisting of a stepped transition from black through the gray range to white.

Green Thumb A viewdata system for farmers established and run by the U.S. Department of Agriculture.

Grid Modulation Modulation produced by introduction of the modulating signal into any of the grid circuits of any tube in which carrier frequency wave is present.

Gridtronics A program service concept which would enable subscribers to individually select the programming delivered to their television set.

Ground (1) The point of reference in an electrical circuit; considered to be at nominal zero potential when other potentials within the circuit are compared to it; (2) Earth connection.

Ground Communication Equipment Satellite earth station electronic equipment.

Ground Rod A copper clad steel or galvanized rod usually eight feet long driven into the ground at the base of a pole or house drop for grounding purposes.

Ground Wire A conductor which provides an electrical connection between an electrical system and earth ground.

Group Delay In the propagation of electromagnetic signals consisting of several frequencies, the difference in propagation transmission time between the highest and lowest frequencies through a device or circuit.

Guard Arm A wooden device mounted directly above and parallel to the strand to indicate a given distance below power company attachments.

Guard Band A frequency band between two channels left unused so as to prevent interference between those channels.

H

Half Duplex A circuit that permits transmission of a signal in two directions, but not at the same time. Contrast with "full duplex" and "simplex."

Halo A dark area surrounding an unusually bright object or a white area surrounding a dark object.

Ham An amateur radio operator.

Hand-helds (1) Pocket-size, microprocessor-based devices often capable of storing as well as displaying data, which can be entered manually (via keyboard), by prerecorded memory, or by bulk transfer over telephone lines or similar transmission media. Examples of hand-helds include electronic games, language translators, and pocket terminals; (2) Portable, battery-operated two-way radios, e.g., "walkie-talkies"; (3) Remote controls for television sets, converters, videocassette recorders, stereos.

Handshaking Exchange of predetermined signals when a connection is established between two data-set devices.

Hang Up An unanticipated stop in a program sequence, caused by a program error.

Hard Copy (1) Any physical document; (2) Computer printout on permanent media such as paper; (3) Facsimile printout on permanent media such as paper.

Hard Disk Disk storage that uses rigid disks rather than flexible disks (floppies) as the storage medium. Hard-disk devices can generally store more information and access it faster. Cost considerations, however, currently restrict their usage to medium and large-scale applications.

Hard Sectoring The use of hardware to define sectors on a disk.

Hard-Wired The direct local wiring of a terminal to a computer system.

Hardware (1) The components that are used to attach aerial cable plant to utility poles, e.g., suspension clamps, through bolts, etc.; (2) Collectively, electronic circuits, components, and associated fittings and attachments; (3) The physical parts, components and machinery associated with computation.

Harmonic Distortion (1) The generation of harmonics by the circuit or device with which the signal is processed; (2) Unwanted harmonic components of a signal.

Harmonically Related Carriers (HRC) A cable plan where each video carrier is a perfect multiple of 6 MHz. This technique is used to mask composite triple beat distortion by zero-beating those distortions with the video carriers.

Harsh A qualifying adjective to describe strident or unmusical sound.

HDTV See "High Definition Television."

Headend The control center of a cable television system, where incoming signals are amplified, converted, processed and combined into a common cable along with any origination cablecasting, for transmission to subscribers. System usually includes antennas, preamplifiers, frequency converters, demodulators, modulators, processors and other related equipment.

Headend Facility The tower, antennas and building housing the headend equipment.

Head Switching Transients White or black transients which occur regularly in band.

Height The size of a picture in the vertical direction.

Helical Recording Format A recording format in which the tape is unwrapped around a cylindrical scanning assembly with one or more recording heads.

Herringbone An interference pattern in a television picture, appearing as either moving or stationary rows of parallel diagonal or sloping lines superimposed on the picture information.

Hertz (Hz) A unit of frequency equivalent to one cycle per second.

Heterodyne To mix two frequencies together in a nonlinear component in order to produce two other frequencies equal to the sum and difference of the

first two. For example, heterodyning a 100-kHz and a 10-kHz signal will produce a 110-kHz (sum frequency) and a 90-kHz (difference frequency) signal in addition to the two original frequencies. Synonymous with beat.

Heterodyne Processor An electronic device used in cable headends which downconverts an incoming signal to an intermediate frequency for filtering, signal level control, and other processing, then reconverts that signal to a desired output frequency.

Hexadecimal An alphanumeric, base-16 system of number notation commonly used in machine language computer programming.

Hi-OVIS See "Higashi-Ikoma Optical Visual Information System."

Hierarchical Network A network in which processing and control functions are performed at several levels by computers specially suited for the functions performed; for example, in factory or laboratory automation.

Higashi-Ikoma Optical Visual Information System (Hi-OVIS) A fiber optic two-way CATV system used in Japan.

High Band That portion of the electromagnetic spectrum from 174 to 216 MHz, where television channels 7 through 13 are located.

High Definition Television (HDTV) A very high quality television signal with picture resolution nearly equal to that of film.

High-Level Language Computer language that allows the programmer to write software programs using verbs, symbols, and commands rather than a machine code. Some common high-level languages are ALCOL, APL, BASIC, COBOL, FORTRAN, PL/I, PL/M, and SNOBOL.

High Level Modulation Modulation produced at the plate circuit of the last radio stage of the system. This generally is AM broadcasting.

High Pass Pertaining to the performance of a circuit that permits the passage of high-frequency signals and attenuates low-frequency signals.

High-Pass Filter A filter which passes frequencies above a given frequency and attenuates all others.

High Power Amplifier A device which provides the energy for carrier amplification necessary to transmit to the satellite.

High Split A cable-based communications system that enables signals to travel in two directions, forward and reverse simultaneously with upstream (reverse) transmission from 5 MHz to about 174 MHz, and downstream (forward) transmission above 230 MHz. Exact crossover frequencies vary from manufacturer to manufacturer.

Highlight Tearing Polarity changes in highlight picture areas. Appears as streaking from white peaks.

Highlights The brightest portions of a picture.

Hit (1) A distinctive sound of very short duration heard from a sound monitor; (2) The process of sending a command to an addressable set-top device.

Homes Passed The number of living units (single residential homes, apartments, condominium units) passed by cable television distribution facilities in a given cable system service area.

Homogeneous Network A network of similar host computers such as those of one model of one manufacturer.

Horizontal Back Porch The interval between the end of horizontal sync and the beginning of the next line of active video, normally about five milliseconds in duration. See also "Back Porch."

Horizontal Bars Thick horizontal bars, alternately dark and light, which extend over the entire picture.

Horizontal Blanking The blanking signal at the end of each scanning line that permits the return of the electron beam from the right to the left side of the raster after the scanning of one line.

Horizontal Displacement A picture condition in which the scanning lines start at relatively different points during the horizontal scan. See also "Serration" and "Jitter."

Horizontal Front Porch That portion of the blanking wave between the end of each video scanning line and the start of horizontal sync. See also "Front Porch."

Horizontal Resolution The maximum number of black and white vertical lines that can be resolved within a horizontal expanse of raster equal to one picture height. NTSC television pictures normally have 300 lines of resolution or less.

Horizontal Retrace The return of the electron beam from the right to the left side of the raster after the scanning of one line.

Host In a packet switching network, the master collection of hardware and software that makes use of packet switching to support user-to-user communications, distributed data processing, and other services.

Host Computer See "Central Computer."

House Drop Coaxial cable from subscriber tap at the utility pole or pedestal to a subscriber's television set.

House Hook A screw device for attaching drop wire or cable to wood frame.

HRC See "Harmonically Related Carriers."

Hub A signal distribution point for part of an overall system. Larger cable systems are often served by multiple hub sites, with each hub in turn linked to the main headend with a transportation link such as fiber optics, coaxial supertrunk, or microwave.

Hub Network A modified tree network in which signals are transmitted to subordinate distribution points (hubs) from which the signals are further distributed to subscribers. See also "Tree Network."

Hue The attribute of color perception that determines whether the color is red, yellow, green, blue, purple, etc.

Hue Shift Banding Banding made visible by hue shifts within a band. See also "Banding" and "Color Banding."

Hum A low-pitched undesired tone or tones, consisting of fundamental and/or several harmonically related frequencies. See also "Hum Modulation."

Hum Bars See "Shutter Bar."

Hum Modulation Undesired modulation of the television visual carrier by power line frequencies or their harmonics (e.g., 60 or 120 Hz), or other low frequency disturbances.

Hyperband The band of cable television channels above 300 MHz.

Hysteresis The difference between the response of a unit or system to an increasing and a decreasing signal.

Hz See "Hertz."

I

IC See "Integrated Circuit."

ICC/IRC See "Incremental Coherent Carriers."

ICS See "Integrated Communication System."

IEEE See "Institute of Electrical and Electronic Engineers."

IF See "Intermediate Frequency."

Image (1) A television picture; (2) A fully processed unit of operational data that is ready to be transmitted to a remote unit; when loaded into the control storage in the remote unit the image determines the operations of the unit; (3) The unwanted product or products of a heterodyne process.

Impact Printer A printer that creates characters by mechanical impact means.

Impairment Scale A scale for the subjective assessment of sound programs and television pictures.

Impedance The combined effect of resistance, inductive reactance, and capacitive reactance on a signal at a particular frequency. In cable television, the nominal impedance of the cable and components is 75 ohms.

Impedance Matching A method used to match two or more components into a single characteristic impedance of one of the components, to minimize attenuation and anomalies.

Impulse Noise Noise characterized by non-overlapping transient disturbances commonly introduced by switches and relays.

Impulse Pay-Per-View (IPPV) The ability to order pay-per-view programming without having to telephone a CATV system office. See also "Pay-Per-View."

Incident Light Reading The foot-candle measurement of light striking the subject.

Incremental Coherent Carriers (ICC/IRC) A cable plan in which all channels except 5 and 6 correspond with the standard channel plan. The technique is used to reduce composite triple beat distortions. Synonymous with incrementally related carriers.

Incrementally Related Carriers See "Incremental Coherent Carriers."

Independent (1) A commercial television broadcast station that generally carries in prime time not more than ten hours of programming per week offered by the three major national television networks; (2) An independently owned and operated cable television system with no MSO affiliation.

Inductor A coil of wire wound, with or without a core of magnetic material, to concentrate the magnetic field and create a higher self-inductance than would be possible with a straight wire.

Informal Order An order or request for service from a broadcaster, which is not transmitted through the normal means of communication with a communications company.

Information Retrieval In digital computer and data processing, the recall upon demand of specific file items.

Infrared That portion of the electromagnetic spectrum just below visible light; infrared radiation has a wavelength from 800 nm to about 1mm. Fiber-optic transmission is predominantly in the near-infrared region, about 800 to 1600 nm.

Ingress The unwanted leakage of interfering signals into a cable television system.

In-House (1) Performing certain tasks with system personnel as opposed to hiring an outside contractor to perform those tasks; (2) Computer system for use only within a particular company or organization.

Initial Error An error represented by the difference between the actual value of a data unit and the value used at the beginning of processing.

Initialization The process carried out at the commencement of a program to test that all indicators and constants are set to prescribed conditions.

Input Converter See "Downconverter."

Input-Output (I/O) See "Radial Transfer."

Insertion Gain A change in signal level expressed in decibels caused by the inclusion of a circuit, circuit section, etc., or item of equipment in a network. See also "Insertion Loss."

Insertion Loss Additional loss in a system when a device such as a directional coupler is inserted; equal to the difference in signal level between input and output of such a device.

Insertion Test Signals See "Vertical Interval Reference Test Signals."

Installation Charge A one-time charge, due upon connection of a new cable television subscriber, used to help recover the actual installation expenditures.

Institute of Electrical and Electronic Engineers (IEEE) An engineering society formed by the merger of the Institute of Radio Engineers and the American Institute of Electrical Engineers.

Institute of Radio Engineers (IRE) Combined with the American Institute of Electrical Engineers January 1, 1963, to form the Institute of Electrical and Electronic Engineers (IEEE).

Institutional System The network of cables of frequencies (upstream) that connect schools, government agencies, etc. to the cable system headend for retransmission downstream to the public through the cable system.

Instructional Program In broadcasting, includes programs involving the discussion of, or primarily designed to further an appreciation or understanding of, literature, music, fine arts, history, geography and the natural and social sciences, as well as programs devoted to occupational and vocational instruction, instruction with respect to hobbies, and similar programs intended primarily to instruct.

Instruction Set The range of commands

that form a programming language.

Instructional Television Fixed Service (ITFS) The ITFS television transmission system was first authorized in 1963 by the Federal Communications Commission for educational television in the 2.5 to 2.686 GHz band. The ITFS band has subsequently been re-allocated for shared operation among multipoint distribution services, multichannel multipoint distribution services, operational fixed services, and ITFS users.

Integrated Circuit (IC) An electronic circuit made by manipulating layers of semiconductive materials.

Integrated Communication System (ICS) A communication system that transmits analog and digital traffic over the same switched network.

Integrated Messenger Cable Coaxial cable with a supporting steel wire or cable embedded into its jacket. See also "Figure 8 Cable."

Integrated System A system in which all components including the various types of amplifiers and taps have been designed from a well-founded overall engineering concept, to be fully compatible with each other.

Intelligent Device A device that, when used with a microprocessor, has data processing ability.

Intelsat See "International Telecommunications Satellite Consortium."

Interactive Pertaining to an application in which each entry calls forth a response from a system or program, as in an inquiry system or an airline reservation system.

Interactive Cable System A two-way cable system that has the capability to provide a subscriber with the ability to enter commands or responses on an in-home terminal and generate responses or stimuli at a remote location. An example of an interactive system would be order entry for Pay-Per-View: the

order information is transmitted upstream on the cable from the subscriber's terminal to the headend, processed by a billing/authorization computer, and authorization to view a specific Pay-Per-View event is sent downstream to the subscriber's terminal.

Intercity Between different urban areas.

Interconnect (1) The process whereby two or more cable operators undertake a joint effort to sell and/or distribute advertising over their CATV systems; (2) The connection of two or more cable systems; (3) The connection of a headend to its hubs.

Interconnection Point Any point in a channel or network where broadcaster and communications company facilities, or two different communications company facilities, are physically connected.

Interface The circuitry that interconnects and provides compatibility between a central processor and peripherals in a computer system.

Intermediate Frequency (IF) In a heterodyne circuit such as that used in a radio receiver, the IF is the frequency that is produced when the frequency of a local oscillator is mixed with the incoming radio frequency (RF) signal.

International Consultative Committee for Radio (CCIR) (Comite Consultatif International des Radio Communications) A committee of the International Telecommunication Union (ITU), Geneva, Switzerland.

International Telecommunication Union (ITU) Organization composed of the telecommunications administrations of the participating nations. Focus is the maintenance and extension of international cooperation for improving telecommunications development and applications.

International Telecommunications Satellite Consortium (Intelsat) Organization composed of governments that adhere to the two major international telecommunications agreements. Stated purpose is the "design, development, construction, establishment, maintenance, and operation of the space segment of the global communications satellite system."

International Telegraph and Telephone Consultative Committee (CCITT) (Comité Consultatif International de Telegrafique et Telephonique) This committee has established a system by which certain measurements may be compared. Such systems are called "weighting networks."

Interstitial Programming Programming that is broadcast between regularly scheduled features, (usually on premium program channels), such as promotional materials or short subjects.

Intracity Within the same urban area.

IRC See "Incrementally Related Carriers."

IRE (1) The Institute of Radio Engineers; (2) A unit of video measurement established by the IRE, where 1 IRE unit equals 0.00714 volts peak-to-peak, and 140 IRE units equals 1 volt peak-to-peak.

IRE Roll-Off A specific gain/frequency characteristic of a video circuit. Usually the high frequency components of a video signal suffer loss or roll-off at a greater rate than do the lower frequency components.

IRE Scale An oscilloscope scale in keeping with IRE Standard 50, IRE 23.S1 and the recommendations of the Joint Committee of TV Broadcasters and Manufacturers for Coordination of Video Levels. See also "IRE."

Isolation The design characteristic of a device which minimizes the transmission of signals in one piece of cable or device into another cable or device.

Iterative Operation The repetition of the algorithm for the solution of a set of equations, with successive combinations of initial conditions or other parameters; each successive combination is selected by a subsidiary computation on a predetermined set of iteration rules.

ITFS See "Instructional Television Fixed Service."

ITS Insertion test signals. See "Vertical Interval Reference Test Signal."

ITU See "International Telecommunication Union."

J-K

Jack A connecting device to which a wire or wires of a circuit may be attached and which is arranged for the insertion of a plug.

Jack Panel A series of jacks arranged and wired in such a way as to provide an easy means of connecting or reconfiguring the overall system.

Jacketed Cable Coaxial cable with a protective covering over the outermost shield.

Jamming Transmitting an interfering signal so as to cause intentional reception impairment.

Jitter An unsteady television picture usually caused by the following: (1) improper synchronizing of lines, groups of lines or entire fields; (2) improper positioning of a film frame with references to the preceding frame in the gate of n or film camera equipment; (3) improper damping in a video tape machine.

Joint Use Simultaneous use of a pole or trench by two or more kinds of utilities.

Jumper Cable (1) Short length of flexible coaxial cable used in older CATV systems to connect the coaxial cable to amplifiers or other CATV components; (2) Short length of coaxial cable used to connect the converter to the subscriber's television set; (3) Any short length of cable or wire generally used to make a less than permanent connection.

K Factor (1) A rating factor given to television transmission and reproducing systems to express the degree of subjective impairment of the television picture; (2) In microwave communications, an index of atmospheric refractivity and effective earth curvature.

Kelvin The temperature unit used with reference to the color temperature of a light source. Zero K is equal to -273.15C.

Key Light The apparent principal source of directional illumination falling upon a subject or area.

Keyboard An alphanumeric, input/output, peripheral device used to communicate with a computer.

Kilobaud The measure of data transmission speed; a thousand bits per second.

Kilobyte A unit of measurement equal to 1024 bytes.

Kine Recording The technique of converting a video image to motion picture film.

Ku Band (1) The group of microwave frequencies from 12 to 18 GHz; (2) The band of satellite downlink frequencies from 11.7 to 12.2 GHz.

L

Lagging Chrominance A picture impairment that occurs when the chrominance portion of a video signal lags the luminance signal, resulting in colors that appear to the right of the image. Synonymous with funny paper effect.

Lagging N The n signal lags the luminance signal. Colors will appear to the right of the image.

Large-Scale Integration (LSI) The process of engraving many thousands of electrical circuits on a small chip of silicon.

Laser Light Amplification by Stimulated Emission of Radiation. A device for generating coherent electromagnetic signals (e.g., light). Low powered lasers are frequently used to transmit light signals into optical fibers.

Lasher A machine designed to spinlash coaxial cable to the supporting messenger strand using a stainless steel lashing wire.

Lashing Wire Clamp A clamping device used for a permanent termination of lashing wire near the pole.

Last In-First Out (LIFO) Processing arrangement used in data manipulation wherein the most recent data input is the first output.

Last In-Last Out (LILO) Processing arrangement used in data manipulation wherein the most recent data input is the last output.

Last Radio Stage The oscillator or radio-frequency-power amplifier stage which supplies power to the antenna. Usually refers to an AM station.

Late Finish The completion of a transmission of program material at some time after the end of the period scheduled for that program.

Late Start The commencement of the transmission of program material at some time after the scheduled starting time.

LCD See "Liquid Crystal Display."

LDS See "Local Distribution Service."

Leading Blacks See "Edge Effect."

Leading Chrominance A picture impairment that occurs when the chrominance portion of a video signal leads the luminance signal, resulting in colors that appear to the left of the image. Synonymous with funny paper effect.

Leading N The n signal leads the luminance signal. Colors will appear to the left of the image.

Leading Whites See "Edge Effect."

Leakage Undesired emission of signals out of a cable television system, generally through cracks in the cable, corroded or loose connections, or loose device closures. Synonymous with signal leakage.

Leapfrogging Carrying a potentially more popular distant similar broadcast signal on a cable system instead of a closer one to provide more appeal and diversity on a cable system.

Lease Back The installation and maintenance of a cable system by a utility such as a telephone company and the subsequent leasing of that system to another company (e.g., a cable company) to operate.

Leased Access Channels Cable television channels specifically designated for leased access services.

Leased Line A line furnished to a subscriber for his exclusive use. Synonymous with private line and dedicated line.

LED See "Light Emitting Diode."

Left (or Right) Signal The electrical output of a microphone or combination of microphones placed so as to convey the intensity, time and location of sounds originating predominantly to the listener's left (or right) of the center of the performing area. Usually refers to stereo broadcasting.

Legally Qualified Candidate Any person who has publicly announced candidacy for nomination by a political party or for nomination or election in a primary, special or general election, municipal, county, state or national, and who meets the qualifications prescribed by the applicable laws.

Level The apparent signal amplitude as indicated on a standard measuring scale.

LIFO See "Last In-First Out."

Light Emitting Diode (LED) A semiconductor which emits light when a proper voltage is applied to its terminals.

LILO See "Last In-Last Out."

Line Extender Feeder line amplifiers used to boost signal and thereby extend the useful range of the feeder cable.

Line Frequency (1) The number of times per second that the scanning spot crosses a fixed vertical line in one direction; (2) Related to commercial power line frequency, i.e., 60 Hz. (3) The horizontal scanning rate of a video signal. For NTSC video, 15.734 KHz.

Line-of-Sight Region The zone between an antenna and all other points without obstruction at the intended operating frequency.

Line Terminator A device used to electrically terminate the end of a coaxial line in its normal impedance, for the primary purpose of minimizing ghosting.

Line-Time Distortion The linear waveform distortion of video signals from 1 to 64 microseconds.

Line-Up Period The period of time used by communications companies (carriers) for verification of the technical and operating parameters of a circuit prior to handover to a customer.

Linear Distortion Distortion resulting from a channel having a linear filter characteristic different from an ideal linear low-pass or band-pass filter; in particular, amplitude characteristics that are not flat over the pass band and phase characteristics that are not linear over the pass band.

Linear Wave-Form Distortion The distortion of the shape of a television waveform signal where the distortion is independent of the amplitude of the signal.

Lineman A person who builds, repairs, and maintains the outside plant of a cable system, telephone company, or power company.

Lines Per Minute (LPM) A description of the speed of a line printer.

Link (1)A successful path between two communications facilities; (2) In computing, a branch instruction, or an address in such an instruction, used to leave a subroutine to return to some point in the main program.

Lip Sync Synchronization of the sound portion with the visual portion of a television program.

Liquid Crystal Display (LCD) A method of creating alphanumeric displays by reflecting light on a special crystalline substance. Frequently used in electronic games and watches, and in portable electronic instruments.

LNA See "Low Noise Amplifier."

LNB/LNC See "Low Noise Block Converter."

Local Area Network (LAN) A system linking together computers, word processors and other electronic office machines to create an inter-office or inter-site network. These networks can also provide access to internal networks, for example, public telephone and data transmission networks, information retrieval systems, etc.

Local Automatic Message Accounting A process using equipment located in a local office for automatically recording billing data for message rate calls (bulk billing) and for customer-dialed station-to-station toll calls.

Local Central Office A telephone central office arranged for terminating subscriber lines and provided with trunks of establishing connections to and from other central offices.

Local Channel A television broadcast station within or close to the cable television service area.

Local Distribution Service (LDS) A fixed community antenna relay station used within a cable television system or systems for the transmission of television signals and related audio signals, signals of standard and FM broadcast stations, signals of instructional television fixed stations, and cablecasting from a local transmission point to one or more receiving points, from which the communications are distributed to the public by cable.

Local Exchange An exchange where telephone subscribers' lines connect.

Local Government Access Channel A cable television channel specifically designated for use by local government.

Local Loop (1) A communication line connecting several terminals in the region of a single controller; (2) That part of a communication circuit between the telephone customer's location and the nearest central office.

Local Origination Programming (LOP) Programs that are produced by the CATV operator rather than those received from television broadcast stations or pay channel distributors. Usually refers to full active video/audio rather than alphanumeric-only material.

Local Oscillator An oscillator, built into the design of the equipment, which generates a signal used in the heterodyne process to mix with incoming signals and produce an intermediate frequency.

Local Signals Television signals received at locations within those stations' predicted Grade B contours.

Log (1) A continuous record of communications kept by a station, or record of the operation of equipment; (2) Abbreviation of logarithm.

Log-Periodic Antenna A directional antenna in which the size and spacing of the elements increase logarithmically from one end of the antenna to the other.

Longitudinal Recording Format A format wherein the record head is stationary and the writing speed equals the tape velocity. Commonly used in audio recorders.

Loop A set of instructions that may be executed repeatedly while a certain condition prevails. In some implementations, no test is made to discover whether the condition prevails until the loop has been executed once.

LOP See "Local Origination Programming."

Low Band That portion of the electromagnetic spectrum from 54 to 88 MHz, where television channels 2-6 are located.

Low-Frequency Interference Interference effects which occur at low frequency, generally considered as any frequency below 15.734 KHz.

Low Level Modulation Modulation produced in a stage earlier than the final stage. Usually refers to broadcasting.

Low Noise Amplifier (LNA) A low noise signal booster used to amplify the weak signals received on a satellite antenna.

Low Noise Block Converter (LNB/LNC) A combination device used on satellite antennas that includes both a low noise amplifier (LNA) to boost the weak signals, and a block downconverter to convert the incoming satellite signals to a lower band of frequencies (e.g., 70-1450 MHz).

Low Pass Filter (LPF) A filter which passes all frequencies below a specified frequency, and blocks those frequencies above the specified frequency.

LPF See "Low Pass Filter."

LPM See "Lines Per Minute."

LSI See "Large Scale Integration."

Lumen Unit of light flux.

Luminance (1) Luminous flux emitted, rejected, or transmitted per unit of solid angle per projected area of the source; (2) The photometric equivalent of brightness; (3) The brightness part of a television picture.

Luminance Signal That portion of the television signal which conveys the luminance or brightness information.

Lux Unit of n equal to 1 lumen per square meter or approximately 0.1 candle power.

M

Machine Language Binary code that can be directly executed by the processor, as opposed to assembly or high-level language.

Magnetic Tape A mylar tape, coated with magnetic particles, on which audio, video or data can be stored.

Main Channel The band of frequencies from 50 to 15,000 Hertz per second which are frequency-modulated to the main carrier. Usually refers to FM broadcasting.

Main Frame (1) The computer itself, that is, the chassis containing the central processor, arithmetic and logic circuits; (2) The central processing unit; (3) Generic term for a computer which is larger than a micro or minicomputer.

Main Memory The principal storage or memory unit in a digital computer or data processing system. Synonymous with main storage.

Main Storage See "Main Memory."

Main Trunk The major cable link or "backbone" from the headend to a community or between communities.

Major Television Market The specified zone of a commercial television station licensed to a community listed in Federal Communications Commission regulations or a combination of such specified zones where more than one community is listed.

Make Good The obligation to re-run a commercial that was miscued, clipped, or for any reason not presented as contracted for by the paying sponsor.

Makeready The cable television preconstruction process performed to ensure adequate clearance from other utilities on aerial installations, verification of easement rights and utility location in underground installations, and the ability of all support structures to withstand the additional loads imposed by the new cables and hardware.

Mandatory Carriage Television signals that, according to FCC regulations, a cable system must carry.

Marker Generator An electronic instrument providing variable or fixed signals and used in conjunction with frequency sweep testing to identify a specific frequency or frequencies in the radio-frequency spectrum.

Mask (1) Pattern of characters used to control the retention or elimination of portions of another pattern of characters; (2) A device located in the front inside of a color picture tube that in combination with the electron gun and colored phosphor dots produces color images.

Masked ROM Regular read-only memory whose contents are produced during manufacture by the usual masking process.

Mass Data A quantity of data larger than the amount storable in a central processing unit of a given computer at any one time.

Mass Memory Large amounts of data usually stored on magnetic disk. Getting data from mass memory is slower than from main memory, but usually the computer does not need to access this information very often.

Mass Storage Device A device having a large storage capacity, for example, a magnetic disk or drum.

Master Antenna Television System (MATV) An antenna and distribution system which serves multiple dwelling complexes such as motels, hotels, and apartments. It is, in effect, a miniature cable system.

Master Console The controlling console that, in a system with multiple consoles, facilitates communication between the operator and the system.

Master Control Program A program that controls the operation of a system, either by connecting subroutines and calling segments into memory as needed, or as a program controlling hardware and limiting the amount of intervention required by an operator.

Master Control Room (MCR) The key location at a network organization center or broadcasting station where overall technical supervisory control and monitoring are accomplished.

Master File A file, used as an authority in a given job, that is relatively permanent, even though its contents may change.

Master/Slave System A system where the central computer has control over, and is connected to, one or more satellite computers.

Matching Transformer An impedance matching device which converts the 75-ohm impedance of the subscriber drop to the 300-ohm impedance of a television or FM receiver.

Matrix In computers, a logic network that forms an array of input leads and output leads with logic elements connected at some of their intersections.

Matrix Printer A printer in which each character is represented by a pattern of dots; for example, a stylus printer, a wire printer.

MATV See "Master Antenna Television System."

Maximum Rated Carrier Power The maximum power at which the radio transmitter can be operated satisfactorily, determined by the design of the transmitter and the type and number of amplifying devices used in the last radio stage.

MCR See "Master Control Room."

MDS See "Multipoint Distribution Service."

MDU See "Multiple Dwelling Units."

Mean Time Between Failure (MTBF) A statistical quantitative value for the time between episodes of equipment or component failure.

Mean Time to Failure The average time a component or system functions without faulting.

Mean Time to Repair The average time required for corrective maintenance.

Media In transmission systems, the structure or path along which the signal is propagated, such as wire pair, coaxial cable, waveguide, optical fiber, or radio path.

Mega (1) Ten to the sixth power, 1,000,000 in decimal notation; (2) When referring to storage capacity, two to the twentieth power, 1,048,576 in decimal notation.

Megabyte A unit of measurement equal to 1024 x 1024 bytes, or 1024 kilobytes; 8 million bits.

MegaHertz (MHz) One million cycles per second.

Memory The section of a computer that "remembers" information, that is, the section that stores and holds data until needed.

Menu A database search method in which the user calls up a table of contents listing specifically numbered subjects. Commonly used in viewdata and teletext systems.

Message Switching A telecommunications technique in which a message is received, stored (usually until the best outgoing line is available), and then retransmitted toward its destination. No direct connection between the incoming and outgoing lines is set up as in line switching. See also "Packet Switching."

Messenger Strand A steel cable, strung between poles or other supporting structures, to which a coaxial cable is lashed and by which it is supported.

Messengered Drop Cable Drop cable with a supporting steel wire embedded into the jacket of the cable.

Metal-Oxide Semiconductor (MOS) A semiconductor developed from new, advanced technology that helped make possible new types of complex chips, such as the microprocessor used in personal computers and high-capacity memories.

Microcomputer A relatively precise term for computers whose central processing units (CPUs) are microprocessor chips. By contrast, mainframes and most minicomputers have CPUs containing large circuitry. Microcomputers include personal computers, small business computers, desktop computers, and home computers. About 400 companies worldwide manufacture microcomputers, which range in price from $200 to $20,000.

Microfiche A system of storing and retrieving information microforms, consisting of film in the form of separate sheets, that contain original text, pictures, data, or anything which has been reduced to micro-images for a greater storage efficiency and arranged in a grid pattern for location of those original images by means of Cartesian coordinates.

Microfilm A system of storing and retrieving information microforms, consisting of film as a data medium, usually in the form of a roll or strip, that contains micro-images of the original information. The images are generally in a sequential arrangement rather than in rows or columns as on microfiche.

Microphonic Bars Light and dark horizontal bars in a television picture which move erratically in a vertical direction due to mechanical shock or vibration.

Microphonics Incidental signal generation caused by mechanical shock or vibration.

Microprocessor A central processing unit implemented on a chip.

Microsecond One millionth of a second.

Microwave A very short wavelength electromagnetic wave, generally above 1000 MHz.

Microwave Antenna Radome A fiberglass cover used to protect an antenna from ice, snow and dirt. Sometimes heated in extreme cold climates.

Midband The band of cable television channels A through I, lying between

120 and 174 MHz.

Midspan A point along the cable and strand in aerial plant between utility poles. A midspan tap is one that is located at a spot between two poles rather than at one of the poles. A midspan drop is one that connects to a subscriber tap at the pole, then runs for some distance along the feeder cable to a span clamp before going to the subscriber's home. Midspan installations are done to avoid crossing property lines or physical obstructions that may prevent a direct aerial run from the pole to the house.

Midsplit System A cable-based communications system that enables signals to travel in two directions, forward and reverse simultaneously with upstream (reverse) transmission from 5 MHz to about 100 MHz and downstream (forward) transmission greater than about 150 MHz. Exact crossover frequencies vary from manufacturer to manufacturer.

Mil A small unit of linear measure; one mil equals 10^{-3} inch (0.001 inch).

Minicomputer An intermediate range computer, between full-size mainframes and 16-bit microcomputers. Historically, minicomputers have served dedicated uses, such as in scientific and laboratory work. The three largest suppliers, Digital Equipment, Hewlett-Packard, and Data General, sell machines ranging in price from $20,000 to $70,000.

Minimum Channel Capacity The minimum number of complete audio/video channels that a given cable system can carry simultaneously.

Minimum Service A minimum number of television signals that, taking television market size into consideration, a cable system may carry.

Mismatch (1) The condition resulting from connecting two circuits or connecting a line to a circuit in which the two impedances are different; (2) Impedance discontinuity.

Mixing Point A designated location at which the contributions of a number of program sources are integrated into a complete program.

MMDS See "Multichannel Multipoint Distribution Service."

Mobile Transmitter A transmitter and antenna system capable of being operated while in motion; for example, a two-way radio or electronic news gathering (ENG).

Mobile Unit An equipped vehicle used to facilitate the production of program material at a location remote from studio facilities.

Mode A method of operation; for example, the binary mode, the interpretive mode, the alphanumeric mode.

Model A representation in mathematical terms of a process, device, or concept.

Modem Modulator/demodulator device, such as is used, for example, to connect a home computer to an ordinary home telephone. See also "Modulation."

Modem-Encryption Devices By placing encryption units at modem interfaces, some systems have all data on the link encrypted and decrypted in a manner that is transparent to the sending and receiving stations.

Modular Constructed with standardized units or dimensions for flexibility and variety in use; allows for easy replacement, substitution, expansion or reconfiguration of modules or subassemblies.

Modulate To vary the amplitude, frequency, or phase of a carrier wave or signal in accordance with the instantaneous amplitude and/or frequency changes of the modulating intelligence.

Modulated Stage The radio frequency stage to which the modulator is coupled and in which the continuous wave (carrier wave) is modulated in accordance with the system of modulation and the characteristics of the modulating wave.

Modulation The process whereby original information can be translated and transferred from one medium to another. Information originally carried as a variation in a particular property (such as amplitude) of one process is transferred and carried as a corresponding variation in some possible different property (such as duration) of the new process.

Moire A wavy or satiny effect produced by convergence of two or more sets of closely spaced lines. A Moire pattern is a natural optical effect when closely spaced lines in the picture are nearly parallel to the scanning lines. In a color television receiver, Moire patterns often produce moving color effects in the noise area.

Monitor (1) A unit of equipment used for the measurement or observation of program material; (2) To observe the picture shading and other factors involved in the transmission of a scene and the accompanying sound; (3) A type of television receiver.

Monochrome Black and white television.

Monochrome Transmission (black and white) The transmission of a signal wave-form which represents the brightness (luminance) values in the picture.

Montage A series of related scenes, sequentially viewed to create a single impression.

MOS See "Metal-Oxide Semiconductor."

MPC See "Multi-Pivoted Coherent Carriers."

MSO See "Multiple System Operator."

MTBF See "Mean Time Between Failure."

MTS Multichannel television sound. See "BTSC."

Multichannel Television Sound (MTS) See "BTSC."

Multidrip (multipoint) A line or circuit interconnecting several stations.

Multichannel Multipoint Distribution Service (MMDS) A collection of various multipoint distribution service (MDS) and instructional television fixed service (ITFS) omnidirectional microwave radio authorizations combined to provide up to 28 channels of entertainment, education and information.

Multi-Outlet Coupler A CATV device which permits serving two or more subscriber television receivers.

Multi-Pay Generic term for two or more premium program services being carried by a subscriber.

Multi-Pivoted Coherent Carriers (MPC) A cable channelization plan to permit use of midband in older cable systems.

Multiple Cable System A system using two or more cables in parallel to increase the information carrying capacity. Synonymous with dual cable system.

Multiple Channel Broadband in nature, that is, capable of carrying several television channels simultaneously.

Multiple Destination Sound Program Transmission A sound program transmission which is simultaneously received by more than one country.

Multiple Dwelling Units (MDU) Residences, passed by cable television lines, with more than one potential customer per building (e.g., apartments or condominiums).

Multiple System Operator (MSO) An organization that operates more than one cable television system.

Multiplex Transmission The simultaneous carrying of two or more signals over a common transmission path.

Multiplexer (MUX) (1) A device or circuit used for mixing signals. In television, multiplexers have several applications; (2) An optical device for combining two picture sources such as film and slides.

Multiplexing In data transmission, a function that permits two or more data sources to share a common transmission medium such that each data source has its own channel. The division of a transmission facility into two or more channels either by splitting the frequency bands, each of which is used to constitute a distinct channel (frequency-division multiplexing), or by allotting this common channel to several different information channels, one at a time (time-division multiplexing).

Multipoint Distribution Service (MDS) A U.S. omnidirectional common carrier microwave radio service authorized to transmit television signals and other communications. MDS operates in the 2150-2162 MHz frequency range, with an effective radius of 30 miles. The service has proved to be an effective means of delivering up to two pay-television programming channels, especially to apartment buildings and hotels. It is not authorized in Canada.

Multiprocessor A computer employing two or more processing units under integrated control.

Multitap A passive device installed in cable system feeder lines to provide signal to the subscriber's drop. A multitap is a combination device which contains a directional coupler that has a hybrid splitter connected to its tap port. Synonymous with directional tap and tap.

Multitasking Pertaining to the concurrent execution of two or more tasks by a computer.

Must Carry See "Mandatory Carriage."

MUX See "Multiplexer."

N

N +1 Created by the Federal Communications Commission, this formula forms the basis by which the FCC regulates expansion of channel capacity for non-broadcast use. The FCC requires that if the government, education, public access, and leased channels are in use at least 80 percent of the Monday-through-Friday period for at least 80 percent of the time during any three-hour period for six consecutive weeks, then within six months the system's channel capacity must be expanded by the operator.

NAB See "National Association of Broadcasters."

NAMIC See "National Association of Minorities in Cable."

Nanosecond (nsec) One billionth of a second (10^{-9}).

Narrowband A relative term referring to a system which carries a narrow frequency range (sometimes used to refer to frequency bandwidths below 1 MHz). In a telephone/television context, telephone would be considered narrowband (3 KHz) and television would be considered broadband (6 MHz).

Narrowcast Channel Group A group of channels where the access conditions may be independently controlled by the computer center on an individual basis. A group contains one or more channels.

Narrowcasting Transmission of information by electromagnetic means, intended for a particularly identified audience (for example, industrial television, special audience cable television, and business and professional programming).

National Association of Broadcasters (NAB) A trade association for the broadcasting (radio, television stations, and television networks) industry.

National Association of FM Broadcasters A nonprofit organization devoted to the promotion and development of FM radio.

National Association of Minorities In Cable (NAMIC) A professional society that serves the needs of minorities in the CATV industry.

National Cable Television Association (NCTA) Washington, D.C.-based trade association for the cable television industry; members are cable television system operators; associate members include cable hardware and program suppliers and distributors, law and brokerage firms, and financial institutions. NCTA represents the cable television industry before state and federal policymakers and legislators. Name was changed in 1969 from National Community Television Association.

National Educational Television (NET) See "Educational Broadcasting Corporation."

National Electric Code (NEC) Safety regulations and procedures issued by the National Fire Protection Association for the installation of electrical wiring and equipment in the United States.

National Electrical Safety Code (NESC) Safety regulations and procedures issued by the American National Standards Institute (ANSI) for the practical safeguarding of persons during the installation, operation, and maintenance of electric supply and communications lines and their associated equipment.

National Electronics Distributors Association (NEDA) Organization whose members are wholesale distributors of electronic components and computer products.

National Sound Program Circuit A sound program circuit which originates and terminates within one country.

National Television Circuit A television circuit which originates and terminates within one country.

National Television System Committee (NTSC) Developers of a color television system which has been adopted by the United States and a number of other countries as their national standard.

NCP See "Network Control Program."

NCTA See "National Cable Television Association."

NEC See "National Electrical Code."

NEDA See "National Electronics Distributors Association."

Negative Image Refers to a picture signal having a polarity which is opposite to normal polarity and which results in a picture in which the white areas appear as black and vice versa.

Negative Transmission Transmission wherein a decrease in initial light intensity causes an increase in the transmitted power.

NESC See "National Electrical Safety Code."

NET National educational television. See "Educational Broadcasting Corporation."

Net Weekly Circulation The estimated number of different television households viewing a particular station at least once per week.

Network (1) A national, regional or provincial organization that distributes programs to broadcasting stations or cable television systems, generally by interconnection facilities. American Broadcasting Company (ABC), Columbia Broadcasting Company (CBS), and National Broadcasting Company (NBC) are the three major broadcasting networks in the United States. (2) A circuit arrangement of electronic components; (3) An interconnected or interrelated group of nodes.

Network Architecture A set of design principles, including the organization of functions and the description of data formats and procedures, used as the basis for design and implementation of a user-application network.

Network Control Program (NCP) A program within the software of a data processing system which deals with the control of the telecommunications network. Normally, it manages the allocation, use, and diagnosis of performance of all lines in the network and of the availability of the terminals at the ends of the network. NCP is also used as a specific term referring to a component of systems network architecture.

Network Cue A predetermined form of signals containing visual and/or audio information employed to program status information.

Network Program Any program furnished by a network (national, regional or special).

Network Programming The programming supplied by a national or regional television network, commercial or noncommercial.

Network Transmission Committee (NTC) A committee of the Video Transmission Engineering Advisory Committee (a joint committee of television network broadcasters and the Bell System) that established technical performance objectives for video facilities leased by the major television networks from the Bell System. The engineering report defining the transmission parameters, test signals, measurement methods, and performance objectives is commonly referred to as NTC-7.

New Build (1) Extension of an existing cable system into a new development or subdivision; (2) The portion of a cable system that is under construction or almost operational; (3) A cable system or portion of a cable system whose construction is planned.

News Program A program that includes reports dealing with current local, national and international events including weather and stock market reports, and, frequently, commentary, analysis and sports news.

Nielsen See "A.C. Nielsen."

Nielsen Diary The diaries used by preselected viewers for the A.C. Nielsen Company wherein the viewers record their daily viewing habits, thus enabling the company to measure television program audience size.

Night Time When applied to broadcasting, that period between local sunset and 12:00 midnight local standard time.

Node For networks, a branching or exchange point.

Noise Random burst of electrical energy or interference which may produce a "salt-and-pepper" pattern over a television picture. Heavy noise is sometimes called "snow."

Noise Factor Ratio of input signal-to-noise ratio to output signal-to-noise ratio.

Noise Figure The amount of noise added by signal-handling equipment (e.g., an amplifier) to the noise existing at its input, usually expressed in decibels.

Noise Immunity The degree to which a circuit or device is not sensitive to extraneous energy, especially noise.

Noise Temperature The temperature that corresponds to a given noise level from all sources, including thermal noise, source noise and induced noise.

Non-Composite Video Signal A signal which contains only the picture signal and the blanking pulses.

Non-Duplication The providing of program exclusivity to a local television broadcast station by refraining from simultaneously duplicating any network program as a result of carrying a similar broadcast station from another (more distant) community on another cable channel.

Nonexclusive Franchise A franchise that allows construction and operation of more than one cable television system within the bounds of the franchise's governmental authority.

Normal Direction The direction of transmission of a signal as specified by contract, agreement or formal order.

Notch Filter A filter which has very high signal attenuation at the frequency to which it is tuned, while passing all other frequencies with minimum attenuation.

Notification Requirements Federal Communications Commission-imposed requirements regarding notice from the CATV operator to television broadcasters, translator stations, superintendents of schools and the FCC prior to beginning operations or supplying subscribers television broadcast signals from distant stations.

Nsec See "Nanosecond."

NTC See "Network Transmission Committee."

NTC-7 A written guideline of video performance measurement procedures and objectives established by the Network Transmission Committee of the Video Transmission Engineering Advisory Committee.

NTSC See "National Television System Committee."

NTSC Video Signal A 525-line color video signal whose frequency spectrum extends from 30 Hz to 4.2 MHz. NTSC video consists of 525 interlaced lines, with a horizontal scanning rate of 15,734 Hz, and a vertical (field) rate of 59.94 Hz. A color subcarrier at 3.579545 MHz contains color hue (phase) and saturation (amplitude) information.

Numeric Database A database primarily containing numbers. It may be used, in conjunction with the appropriate software, for various types of analyses or report generation.

O

Object Code Output from a compiler or assembler which is itself executable code or is suitable for processing to produce executable machine code.

Occasional Circuit/Link/Connection A circuit, link or connection set up between two stations on an as-required basis.

Occasional Service Service performed or facilities supplied on a per-occasion basis for a limited duration of time.

OCR See "Optical Character Recognition."

OEM Original equipment manufacturer.

Off Air See "Off-the-Air."

Off-Line Mode of operation in which terminals, or other equipment, can operate while disconnected from a central processor. Contrast with "On-line."

Off Network Series A series whose episodes have had a national network television exhibition in the United States or a regional network exhibition in the relevant market.

Off-the-Air Refers to the reception of signals broadcast directly through the air through the use of a local antenna.

Synonymous with off air and over-the-air.

Office of Telecommunications Policy (OTP) A division of the President's Executive Office, the OTP is responsible for researching and evaluating policy questions, developing legislative proposals, and advising the Executive Branch on all areas of communications policy.

On-Line Indicating direct connection to a host computer.

On-Line Billing The ability to immediately input and output data (such as changes in levels of service) regarding subscriber billing information.

Open Loop System A servo correction system in which the residual error is unrelated to the means of correction. Compare with "Closed Loop System."

Operating Cash Flow Total revenues less operating, general, and administrative expenses before depreciation and amortization expenses.

Operating Income The difference between all revenues received and operating, depreciation, and other expenses and taxes.

Operating Permit Authorization by a non-municipal government entity which allows the construction and operation of a cable television system within the bounds of its governmental authority.

Operating Power The power actually supplied to the radio station antenna. Refers to a broadcast operation.

Operational Communications Communications related to the technical operation of a broadcasting station and its auxiliaries, other than the transmission of program material and cues and orders directly concerned therewith.

Optical Character Recognition (OCR) The machine identification of printed characters through use of light-sensitive devices; often used as a method of entering data.

Optical Fiber An extremely thin, flexible thread of pure glass able to carry one thousand times the information possible with traditional copper wire.

Order Wire See "Engineering Service Circuit."

Ordinance Municipal or local law and/or regulations passed to establish guidelines for the cable franchising process.

Origination Cablecasting Programming (exclusive of broadcast signals) carried on a cable television system over one or more channels and subject to the exclusive control of the cable operator.

Oscilloscope An oscillograph test apparatus primarily intended to visually represent test or trouble-shooting voltages with respect to time. Synonymous with scope.

OTP See "Office of Telecommunications Policy."

Outlet (75 Ohm) A CATV connection terminal which connects a television receiver through a 75-ohm coaxial cable, using a 75- to 300-ohm matching transformer at the receiver terminals to the cable system. The outlet is usually a wall plate mounted near the television set.

Outlet (300 Ohm) A CATV outlet designed to connect directly to the antenna terminals of a television or FM receiver. Now obsolete due to Federal Communications Commission signal leakage rules.

Output Converter An electronic device which upconverts an intermediate frequency to a desired frequency. Generally the output stage of a headend modulator or processor. Synonymous with upconverter.

Output Level The signal amplitude, usually expressed in decibel millivolts, at the output port of an active or passive device.

Output Tilt See "Tilt."

Outside Broadcast A collective term including remote pickup or remote pickups, the program material contributed by each and the coordination and control required to blend all into one program.

Over-the-Air See "Off-the-Air."

Overbuild The construction of a second, competing CATV system in a franchise area already served by a CATV company.

Overhead Equipment placed above ground on supporting structures. See also "Aerial."

Overload-to-Noise Ratio The ratio of overload-to-noise level measured or referred to at the same point in a system or amplifier, usually expressed in decibels, and commonly used as an amplifier figure of merit or performance specification.

Overshoot An excessive response to an unidirectional voltage change.

P

PABX See "Private Automatic Branch Exchange."

Packet An addressed data unit of convenient size for transmission through a network. See also "Packet Switching."

Packet Switching A type of data communications in which small defined blocks of data, called packets, are independently transmitted from point to point between source and destination, and reassembled into proper sequence at the destination.

Packing Density The number of storage cells per unit length, unit area, or unit volume; for example, the number of bits per inch stored on a magnetic tape track or magnetic drum track.

Page (1) A display of data on a CRT terminal which fills the screen; (2) A unit of viewdata information which may consist of one or more frames; (3) In a virtual storage computer system, a fixed-length block that has a virtual address and that is transferred as a unit between real storage and auxiliary storage.

Pagination In word processing, the process in which a word processor adjusts a multipage document as it is revised in order to ensure uniform page length and appearance.

Pairing A partial or complete failure of interlace in which the scanning lines of alternate fields do not fall exactly between one another but tend to fall (in pairs) one on top of the other.

PAL See "Phase Alteration Line."

Pan To move the camera slowly left or right. Term more rarely used in up and down motions.

Parabolic Antenna An antenna that has a folded dipole or feed horn mounted at the focal point of a metal, mesh, or fiberglass dish having a concave shape known as a parabola.

Parabolic Dish In radio communications, an antenna whose cross-section is a parabola.

Parallel Input/Output Inputting data to, or outputting data from, storage in whole information elements, e.g., a word rather than a bit at a time. Typically, each bit of a word has its own wire for data transmission, so that all of the bits of a word can be transmitted simultaneously.

Parity Check A check of the accuracy of data being transmitted. To accomplish this, an extra parity bit is added to a group of bits so that the number of ones in the group is, according to the specification, even or odd. Then, at the receiving end, the bits in the word are added, the parity bit needed for that total is determined, and the total is then compared with the parity bit transmitted.

Partial Network Station A commercial television broadcast station that generally carries in prime time more than ten hours per week offered by the three major national television networks, but less than that programmed by a full network station.

Pass Band The range of frequencies passed by a filter, amplifier or electrical circuit.

Passive Device A device basically static in operation, that is, it is not capable of amplification or oscillation, and requires no power for its intended function. Examples include splitters, directional couplers, taps and attenuators.

Passive Repeater A reflecting device used to redirect microwave energy without adding radio frequency energy.

Password A group of characters, letters, or numbers which must be entered and verified by the computer for an individual to have access to a computer system or program.

Pay Cable Pay television programs distributed on a cable television system and paid for by an additional charge above the monthly cable subscription fee. Fee may be levied on several bases: per program, per channel, per tier, etc. See also "Pay Television."

Pay-Per-View (PPV) Usage-based fee structure used sometimes in cable television programming in which the user is charged a price for individual programs requested.

Pay Television A system of distributing premium television programming either over the air, or by cable, for which the subscriber pays a fee. The signals for such programming may be scrambled to keep non-subscribers from receiving service. A decoder or descrambler might be used to allow the paying subscribers to receive the pay television programming. Synonymous with premium television.

PC Board See "Printed Circuit (PC) Board."

PCM See "Pulse Code Modulation."

PE See "Polyethylene."

Peak Load A higher-than-average quantity of communications traffic; usually expressed for a one-hour period and as any of several functions of the observing interval, such as peak hour during a day, average daily peak hours over a 20-day interval, maximum of average hourly traffic over a 20-day interval.

Peak Power The power over a radio frequency cycle corresponding in amplitude to synchronizing peaks. Refers to television broadcast transmitters.

Peak Program Meter (PPM) A peak level indicator used in the measurement of speech and music on sound program transmissions.

Peak-to-Peak The amplitude (voltage) difference between the most positive and the most negative excursions (peaks) of an electrical signal.

Pedestal Housing See "Undergound Housing."

Penetration In areas where cable television is available, the percentage of households passed by cable distribution facilities that subscribe to the service. Synonymous with saturation.

People Meter A recording device placed in viewers' homes that allows broadcast and CATV programmers to measure television programs' audience sizes.

Percentage Modulation (Amplitude) The ratio of half the difference between the maximum and minimum amplitudes of an amplitude modulated wave to the average amplitude expressed in percentage.

Percentage Modulation (FM) As applied to frequency modulation, (1) the ratio of the actual frequency swing defined as 100 percent modulation, expressed in percentage; (2) The ratio of half the difference between the maximum and minimum frequencies of the average frequency of an FM signal.

Performance Standards Certain minimum technical requirements, established by the appropriate regulatory entity, which must be met by a cable system operator.

Peripheral Device such as a communications terminal that is external to the system processor.

Permanent Circuit/Link/Connection A circuit, link or connection set up between two stations on a permanent basis.

Personal Computer A low-cost, portable computer with software oriented towards easy, single-user applications.

PERT See "Program Evaluation and Review Technique Network."

Phase A fraction, expressed in degrees, of one complete cycle of a wave form or orbit.

Phase Alteration Line (PAL) The color television standard used in Australia, most of Europe, and the United Kingdom.

Phase Distortion Distortion characterized by input-to-output phase shift between various components of a signal passed by a circuit or device.

Phase Lock Loop An electronic servo system controlling an oscillator so that it maintains a constant phase angle relative to a reference signal source.

Phase Modulation A form of modulation in which the phase of the carrier is varied in accordance with the instan-

taneous value of the modulating signal.

Phase-Shift Keying Modulation technique for transmitting digital information whereby that information is conveyed by selecting discrete phase changes of the carrier.

Phosphorescence Emission of light from a substance during and after excitation has been applied, such as the brightness lines on a television picture tube.

Pickup Tube An electron-beam tube used in a television camera where an electron current or a charge-density image is formed from an optical image and scanned in a predetermined sequence to provide an electrical signal. A device that converts optical images to electrical impulses, as in television picture generation.

Picture Monitor A cathode-ray tube or similar device and its associated circuits, arranged to view a television picture.

Picture Signal That portion of the composite video signal which lies above the blanking level and contains the picture brightness information.

Picture Tube The television cathode-ray tube used to reproduce and display an image created by variations of intensity of the electron beam which scans the coated surface on the tube interior.

Pilot Carrier Signals on CATV systems used to operate attenuation (gain) and frequency response (slope) compensating circuitry in amplifiers.

Pilot Subcarrier A subcarrier serving as a control signal for use in the reception of stereophonic broadcasts.

Pipelining Commencing one instruction sequence prior to the completion of another.

Pirating Signals Reception of a pay television or cable television transmission without authorization.

Pixel The smallest controllable element

that can be illuminated on a display screen. Closely related to resolution.

Pixel Pattern The matrix used in constructing the symbol or character image on a display screen.

Plumbing A generic term which refers to coaxial cables, waveguides, and related hardware.

P.O. Purchase order.

Point-of-Sale Terminal A device which operates as a cash register in addition to transmitting information.

Point-to-Multipoint A wired or wireless communication link between a single fixed location or user (usually the sender) and two or more other fixed locations or users (usually the receivers). Two examples are multipoint distribution service (MDS) and multichannel multipoint distribution service (MMDS).

Point-to-Point Communication A wired or wireless communication link between two specific fixed locations or users, usually, but not always, two-way.

Point-to-Point Connection A connection established between two data stations for data transmission. The connection may include switching facilities.

Polarity of Picture Signal Refers to the polarity of the black portion of the picture signal with respect to the white portion of the picture signal. For example, in a "black negative" picture, the potential corresponding to the black areas of the picture is negative with respect to the potential corresponding to the white areas of the picture, while in a "black positive" picture, the potential corresponding to the black areas of the picture is positive. The signal as observed at broadcasters' master control rooms and telephone company television operating centers is "black negative," usually as seen on an oscilloscope.

Polarization The orientation of the electric field as radiated from the transmitting antenna, usually vertical, horizontal, or circular. Circular can be either left-handed (clockwise) or right-handed (counter-clockwise) sense.

Polarization Diversity A method of diversity transmission and reception in which the same information is transmitted and received simultaneously on orthogonally polarized waves, thus resulting in less signal fading due to propagation anomalies.

Pole Attachment See "Pole Rights."

Pole Contact Point The vertical contact point on each utility pole for the CATV cable.

Pole Hardware Bolts, washers, suspension clamps and other related equipment, including guys, guy anchors, and guy guards used to mechanically attach CATV messengers or brace or structually support or reinforce the utility pole.

Pole Rearrangements The physical movement of telephone and/or power cables or equipment on a utility pole to accommodate CATV cable.

Pole Rights An agreement between the CATV operator and the utility companies or other owners of poles upon which the operator has been granted the right of attaching his hardware for the suspension of his cable.

Polyethylene (PE) A form of dielectric insulation used in coaxial cables.

Polyvinyl Chloride (PVC) A plastic used on some types of coaxial cable as an outer jacket. This material is also used in the manufacture of some types of plastic pipe used as underground conduit.

Port (1) A communication channel between a computer and another device, such as a terminal. The number of ports determines the number of simultaneous users; (2) With regard to electrical devices, the input(s) and output(s).

Portable Transmitter A transmitter so constructed that it may be moved about conveniently from place to place but not ordinarily used while in motion, although some portable communications equipment does provide the capability to be used while in motion.

Post, Telegraphs, and Telephones (PTT) Generic term indicating the various types of communications networks used in countries throughout the world.

Power Doubling An amplification technique where two amplifying devices are operated in parallel to gain an increase in output capability.

Power Down Pre-arranged steps undertaken by a computer or operator when power fails or is shut off in order to preserve the state of the processor or data and minimize damage to peripherals.

Power Gain The amount by which power is increased by the action of an amplifier, usually expressed in decibels.

Power Loss Power dissipated or attenuated in a component or circuit, usually expressed in decibels.

Power Pack (1) An electronic device in an amplifier housing which converts low voltage AC to regulated DC voltages suitable for operating other modules in the housing. See also "Power Supply"; (2) A battery and, usually, its recharging equipment used to power portable cameras, lights and recording or transmitting equipment.

Power Splitter A passive device which divides the input signal into two or more outputs.

Power Supply As used in CATV systems, (1) a stepdown AC transformer which supplies low voltage AC (usually 60 volts) to operate amplifiers in the system; (2) the module located in the amplifier housing that converts the low

voltage AC to regulated DC for actual operation of the electronic devices inside the housing.

PPM See "Peak Program Meter."

PPV See "Pay-Per-View."

Preamplifier A low noise electronic device (usually installed near an antenna) designed to strengthen or boost a weak off-air signal to a level where it will overcome antenna downlead loss and be sufficient to drive succeeding processors or amplifiers.

Predicted Grade B Contour See "Grade B Service."

Pre-Emphasis A change in the relative level (amplitude) of some frequency components of a signal with respect to the other frequency components of the same signal. The high frequency portion of a pre-emphasized band is usually transmitted at a higher level than the low-frequency portion of the band in an attempt to compensate for the greater losses usually suffered at higher frequencies, and the adverse effects of noise at the higher frequencies.

Preparatory Period A period after the start of a service period during which customers (broadcasters) may perform their own tests, adjustments, and other work as necessary.

Preprocessor A computer program that effects some preliminary computation or organization.

Pressure Tap A device (now obsolete) which connects to the center conductor and shield of a distribution cable to extract television signals; a pressure tap does not require complete cutting of cable to make contact for a subscriber drop.

Prestel The British Post Office's public viewdata service.

Primary Service Area The area in which the broadcast ground wave is not subject to objectional interference or objectional fading. Usually refers to AM broadcasting.

Prime Station The television broadcasting station radiating the signals which are retransmitted by a television broadcast translator station, or picked up by a CATV headend.

Prime Time The three-hour period from 8:00 to 11:00 p.m., local time, except in the Central Time Zone, where the relevant period is between the hours of 7:00 and 10:00 p.m., and in the Mountain Time Zone, where each station elects whether the period will be 8:00 to 11:00 p.m. or 7:00 to 10:00 p.m.

Principal City Service Satisfactory service expected for at least 90 percent of the receiving locations. See also "Grade A Service" and "Grade B Service."

Principal Community Contour The signal contour which a television station is required to place over its entire principal community. See also "Principal City Service."

Printed Circuit (PC) Board A circuit board whose electrical connections are made through conductive material that is contained on the board itself, rather than with individual wires.

Printed Dirt A random distribution of white or light gray spots, caused by dirt on the negative, that occurs in the picture reproduced from a motion picture film.

Prism A solid, often triangular-shaped glass, plastic, or crystal transparent device frequently used to optically divide or redirect light direction. In CATV, used mostly in color television cameras.

Private Automatic Branch Exchange (PABX) A private automatic telephone exchange, usually located at the user's site, that routes and interfaces the local business telephones and data circuits to and from the public telephone network.

Private Automatic Exchange A dial telephone exchange, usually located at the user's site, that provides private telephone service to an organization and that does not allow calls to be transmitted to or from the public telephone network.

Private Automatic Switching System A series of packaged PBX service offerings provided on the basis of service features rather than specific hardware.

Private Branch Exchange (PBX) A manual exchange connected to the public telephone network on the user's premises and operated by an attendant supplied by the user.

Private Line Service See "Leased Line."

Private Voice-Band Network A network made up of voice-band circuits and sometimes switching arrangements, for the exclusive use of one customer. These networks can be nationwide in scope and typically serve large corporations or government agencies.

Processing Marks (1) Spots or marks of various shapes and sizes on a film which are caused by defects in the processing or drying of films; (2) Random variations in film density, running longitudinally, due to failure to process uniformly the images on a film.

Processor See "Heterodyne Processor."

Program Evaluation and Review Technique (PERT) Network An extensive study of an overall program to list all individual activities, or jobs, that must be carried out to fulfill the total objective. These efforts are then arranged in a network that displays their relationship.

Program Non-Duplication A regulation that sometimes requires a cable system to black out or remove the programming of a distant station that it carries to avoid duplicating a local station's programming.

Programmable Memory See "Random Access Memory."

Programmable Read-Only Memory (PROM) A type of read-only memory that can be programmed by the computer user. This programming usually requires special equipment.

Programming (1) The designing, writing, and testing of computer programs; (2) The news, entertainment, information resources, and educational presentations carried on a cable system or broadcast by a radio or television station. Because such programming can originate at the local, regional, or national level, it offers a rare opportunity to tailor presentations to the current and future varied needs of the community.

Programming Language An artificial language, established for expressing computer programs, which uses a set of characters and rules whose meanings are assigned prior to use.

Projection Television A system, using a combination of lenses and mirrors, which projects an enlarged television picture on a screen from a special, very bright television picture tube or tubes.

PROM See "Programmable Read-Only Memory."

Prompt Any symbol or message presented to an operator by an operating system, indicating a condition of readiness, location, or that particular information is needed before a program can proceed.

Prop Theatrically derived abbreviation for property; any portable article on the set.

Propagation The act or process of radio waves passing through space or the atmosphere.

Propagation Loss Energy lost by a signal during its passage through the transmission medium.

Proportional Spacing The function whereby characters are spaced according to their natural width.

Protection Channel The broadband channel of a carrier system that is utilized as a spare and can be switched into service in the event of a failure of a normal working broadband channel.

Protocol (1) A specification for the format and relative timing of functional units of a communication system that must be followed if communication is to be achieved; (2) The set of rules governing the operation of functional units of a communication system that must be followed if communication is to be achieved.

Psophometer A noise-measuring set which includes a CCITT weighting network.

Psophometric Noise Level Noise level measured using a qualified psophometer weighting network.

PTT See "Post, Telegraphs and Telephones."

Public Access Channel A cable television channel specifically designated as a noncommercial public access channel available on a first-come, non-discriminatory basis.

Public Affairs Program Includes talks, commentaries, discussions, speeches, editorials, political programs, documentaries, forums, panels, round tables and similar programs primarily concerning local, national and international public affairs.

Public Message Service The public telegram system offered by Western Union.

Public Network A network established and operated by communication common carriers, or telecommunication administrations, for the specific purpose of providing circuit-switched, packet-switched, and leased-circuit services to the public.

Public Service Announcement An announcement for which no charge is made and which promotes programs, activities or services of federal, state or local governments, or the programs, activities, or services of non-profit organizations and other groups regarded as serving community interests.

Public Switched Network Any switching system that provides a circuit switched to many customers. In the United States, there are four: Telex, TWX, telephone, and Broadband Exchange.

Public Telephone Network The traffic network that provides public telephone service.

Public Telephone Service Ordinary telephone service in which a customer has a connection to a central office and can be connected to any other customer of the service. Synonymous with telephone service.

Public Television Noncommercial television broadcasting.

Public Utilities Commission (PUC) Generic term for the state government agency responsible for overseeing utilities and their regulation and performance.

PUC See "Public Utilities Commission."

Pulse A variation in the value of a quantity, short in relation to the time schedule of interest, with the final value being the same as the initial value.

Pulse Amplitude Modulation A modulation technique in which the amplitude of each pulse is related to the amplitude of an analog signal. Used, for example,

in time-division multiplex arrangements in which successive pulses represent samples from the individual voice-band channels; also used in time-division switching systems of small and moderate size.

Pulse and Bar Test Signal A video test signal which contains, on one or more lines, a sine squared pulse and white bar transmitted with synchronizing pulses.

Pulse Code Modulation (PCM) A form of modulation in which the modulating signal is sampled and the sample quantized and coded so that each element of information consists of different kinds of numbers of assigned pulses and spaces that can be converted back to the original signal at the receiving end. PCM systems have inherent security and noise immunity features.

Pulse Rate The time interval of periodic pulses that are integrated with the control of a computer or total system.

Pulse Risetime The time required for the leading edge of a pulse to rise from 10 to 90 percent of its maximum amplitude. Synonymous with risetime.

Push-Button Dialing The use of keys or push buttons instead of a rotary dial to generate a sequence of digits to establish a circuit connection.

PVC See "Polyvinyl Chloride."

Q-R

Quadrapower An amplification technique whereby four output devices (or two power doubling devices) are operating in parallel to increase output capability.

Quadrature Crosstalk Color contamination at color transition resulting from the interaction of the chrominance signal side bands.

Quadrature Error One or more groups of the head band of approximately 16 lines of a video tape playback displayed horizontally as compared to the rest of the picture or other groups of bands.

Quantizer A component of a digital communications system whose function is to assign one of a discrete set of values to the amplitude of each successive sample of a signal. The discrete set of values corresponds to a discrete set of contiguous non-overlapping intervals covering the dynamic amplitude range of the signal.

Quantum Clock A device that allocates an interval or quantum of processing time to a program established by priorities used in computing systems that have time-sharing procedures.

QUBE A two-way cable system operated by Warner Amex in Columbus, Ohio, and introduced in Houston, Cincinnati, and Pittsburgh.

R & D Research and development.

Radial Transfer The process of transmitting data between a peripheral unit and a unit of equipment more centrally located. Synonymous with input-output (IO), input process.

Radiating Element The element of an antenna from which electromagnetic energy is directly radiated.

Radio Frequency (RF) An electromagnetic signal above the audio and below the infrared frequencies.

Random Access A method of providing or achieving access where the time to retrieve data is constant and independent of the location of the item addressed earlier.

Random Access Memory (RAM) A volatile memory used by a computer's central processing unit as a chalkboard for writing and reading information. RAM is measured in multiples of 4096 bytes (4K bytes), and serves as a rough measurement of a computer's capacity.

Most computers have a minimum of 16K bytes, and many personal computers have up to 640K or more. Synonymous with programmable memory.

Random Noise Thermal noise generated from electron motion within resistive elements of electronic equipment.

Raster The scanned (illuminated) area of a television picture tube.

Rating The measurement of television program audience that represents a percentage of total television households watching a particular program.

Raw Data Data that has not been processed in any way.

Read-Only Memory (ROM) A type of permanent, non-erasable memory that plugs directly into the wiring of a computer, and contains computer programs. Some computers are supplied with some built-in ROM, whereas others have external slots for inserting ROM cartridges.

Readout Display of processed information on a terminal screen.

Ready-Access Terminal A class of un-sealed terminals used to make connections of customer drop wires to wire pairs in a distribution cable.

Real-Time Clock An electronic time-keeping circuit within a digital computer that produces periodic signals that reflect the interval between events; can sometimes be used to give the time of day.

Rebroadcast Reception by radio of the programs of a radio station, and the simultaneous or subsequent retransmission of such programs by a broadcast station.

Rebuild The physical upgrade of a cable television system, often involving the replacement of amplifiers, power supplies, passive devices, and sometimes the cable, strand, hardware, and subscriber drops.

REC See "Recorded Program."

Receiver Electronic device which can convert electromagnetic waves into either visual or aural signals, or both. For cable television, usually the subscriber's television set.

Reciprocity Theorem A theorem stating that the directional receiving pattern of an antenna is identical with its directional pattern as a transmitting antenna.

Recorded Program (REC) Any program using recordings, transcriptions or tapes.

Rectifier A device that can convert an alternating current (AC) into a direct current (DC).

Recursive Function A function whose values are natural numbers that are derived from other natural numbers by substitution formulae in which the function is an operand.

Recursive Process A method of computing values of functions where each stage or processing contains all subsequent signs. That is, the first stage is not completed until all other stages are ended.

Reference Black Level The level corresponding to the specified maximum excursion of the luminance signal in the black direction.

Reference Signals (Vertical Interval) (1) Signals inserted into the vertical interval of the program source which are used to establish black and white levels. Such a signal might consist of five microseconds of reference black at 7.5 IRE divisions and five microseconds of reference white at 100 IRE divisions located near the end of lines 18 and 19 of the vertical interval; (2) Reference signals in the vertical interval used for color reference for television receivers, and for on-line testing of television transmission.

Reference White Level The level corresponding to the specified maximum excursion of the luminance signal in the white direction.

Reflection Coefficient At an impedance mismatch, the ratio of the reflected energy to the incident energy. See also "Voltage Standing Wave Ratio (VSWR)" and "Return Loss."

Reflections The full or partial return of transmitted electromagnetic energy to the source by an impedance mismatch.

Relative Burst Amplitude Distortion Distortion in which amplitudes of all color components are changing by an equal amount.

Relative Burst Phase Distortion Distortion in which phases of all color components shift equally.

Relative Chroma Level The difference between the level of the luminance and chrominance signal components.

Relative Chroma Time The difference in absolute time between the luminance and chrominance signal components.

Remote Access Data Processing Communication with a data processing facility through a data link. Synonymous with teleprocessing.

Remote Control Operation Operation of a station by a qualified operator at a control position from which the transmitter is not visible. The control position is equipped with suitable control and telemetering circuits so that the essential functions can be performed from the control point as well as from the transmitter.

Remote Pickup The process of originating program material outside of a permanent broadcast or cable studio building.

Remote Pickup Broadcast Mobile Station A land mobile station licensed for transmission of program material and related communications from the scene of events, which occur outside a studio, to broadcasting stations; can also be used for communicating with other remote pickup broadcast base and mobile stations.

Remote Pickup Broadcast Station Unit made up of a remote pickup broadcast base station and a remote pickup broadcast mobile station.

Remote Pickup Point A location, outside of permanent broadcast studio buildings, which is provided with channels to the mixing point, thus permitting remote origination of program materials.

Remote Pickups Events televised away from the studio by a mobile unit or by permanently installed equipment at the remote location.

Remote Switch A small central office or subsystem of a switching entity that interfaces telephone subscriber lines and concentrates those lines into channels to a remotely located base or host switch.

Repeater Point Premises at which reception, amplifying and associated apparatus is installed to permit adjustment of electrical signal for retransmission.

Request for Proposal (RFP) Issued by a purchasing entity (e.g., a CATV franchising authority), a request for proposal outlines guidelines, specifications, and requirements that must be met by suppliers (e.g., potential system operators) in order to bid on a project (or system).

Resolution A measure of picture resolving capabilities of a television system determined primarily by bandwidth, scan rates and aspect ratio. Relates to fineness of details perceived.

Resolution (Horizontal) The amount of resolvable detail in the horizontal direction in a picture. It is usually expressed as the number of distinct vertical lines, alternately black and white, which can be seen in three-quarters of the width of the picture. This information usually is derived by observation of the vertical wedge of a test pattern. A picture which is sharp and clear and shows small details has good, or high, resolution. If the picture is soft and blurred and small details are indistinct it has poor, or low, resolution. Horizontal resolution depends upon the high frequency amplitude and phase response of the pickup equipment, the transmission medium, and the picture monitor, as well as on the size of the scanning spots.

Resolution (Vertical) The amount of resolvable detail in the vertical direction in a picture. It is usually expressed as the number of distinct horizontal lines, alternately black and white, which can be seen in a test pattern. Vertical resolution is primarily fixed by the number of horizontal scanning lines per frame. Beyond this, vertical resolution depends on the size and shape of the scanning spots of the pickup equipment and picture monitor and does not depend upon the high frequency response or bandwidth of the transmission medium or picture monitor.

Response Time The time interval between the instant a signal or stimulus is applied to or removed from a device or circuit, and the instant the circuit or device responds or acts.

Retrace The return of a scanning beam to a desired position.

Retrofitting The installation of additional equipment or the rebuilding of sections of a cable distribution system after it has been installed. Sometimes needed to increase channel capacity or to provide interactive service.

Return Feed The signal material being sent upstream on a CATV system to the headend from a point out in the cable system.

Return Loss The ratio of input power to reflected power. This measure of impedance dissimilarity is usually expressed in decibels when applied to CATV testing.

Reverse Direction Indicates signal flow direction is toward the headend. Low frequencies are amplified in this direction.

RF See "Radio Frequency."

RFP See "Request for Proposal."

Ring (Network) A network in which each node is connected to two adjacent nodes.

Ringing Picture interference caused by frequency sensitive high-Q circuits such as traps or filters; results when abrupt changes in the video or radio frequency signal level shock excites the circuit into dampened oscillation, or attempts to force signal transients to occur at faster time than the circuit will allow. Ringing appears as repetitive, very close-in ghosting on high contrast ledges.

Risetime See "Pulse Risetime."

RMS See "Root Mean Square."

Roll A television picture which moves up or down due to lack of correct vertical synchronization.

Roll Off A gradual or sharp attenuation of gain versus frequency response at either or both ends of the transmission pass band.

Roll-Off Frequency The frequency at which the gradual or sharp change in gain versus frequency occurs.

Roll-Over A feature of higher-quality keyboards that allows keys to be pressed nearly simultaneously in rapid typing bursts.

ROM See "Read-Only Memory."

Root Mean Square (RMS) (1) The square root of the sum of the squares of the intensities or amplitudes of individual components of a function, such as the frequency components of a signal or of electromagnetic radiation; (2) The square root of the mean of the squares of a set of values.

RS232 Interface A universal interface system to connect data terminals, modems and printers.

S

S Distortion A curving reproduction in a television picture of a vertical straight line.

Safety Cone An orange or red plastic cone, several of which are placed around service trucks and work areas located in or near the flow of vehicular traffic, for the purpose of diverting traffic around the service truck or work area. Synonymous with cone and traffic cone.

Safety Strap A strap that attaches to a lineman's body belt for the purpose of securing the lineman to the utility pole or other structure being climbed.

Sag The vertical drop distance, usually measured at the midpoint of a cable, referenced from an imaginary straight line connecting the two supporting ends of the cable.

SAP See "Second Audio Program."

Satellite An orbiting space station primarily used to relay signals from one point on the earth's surface to one or many other points. A geosynchronous or "stationary" satellite orbits the earth exactly in synchronization with the earth's rotation and can be communicated with using fixed non-steerable antennas located within the satellite's "footprint."

Satellite Business Systems (SBS) A domestic satellite carrier, authorized by the FCC in 1977, and owned by subsidiaries of Comsat General, IBM, and Aetna Casualty and Surety Company. On June 11, 1980, SBS filed with the Federal Communications Commission for authority to provide a private network service, Communications Network Service (CNS), offering voice and data transmission, teleconferencing, and other services at high speeds, using dedicated customer premises-located earth stations shared by CNS customers.

Satellite Earth Terminal That portion of a satellite link which receives, processes and transmits communications between earth and a satellite.

Satellite Master Antenna Television System (SMATV) A system wherein one central antenna is used to receive signals (broadcast or satellite) and deliver them to a concentrated grouping of television sets (such as might be found in apartments, hotels, hospitals, etc.).

Satellite Receiver A microwave receiver capable of receiving satellite transmitted signals, downconverting, and demodulating those signals, and providing a baseband output (e.g., video and audio). Modern receivers are frequency agile, and some are capable of multiple band reception (e.g., C-band and Ku-band).

Satellite Relay An active or passive satellite repeater that relays signals between two earth stations.

Saturation See "Penetration."

Saturation Banding Banding made visible by the difference in saturation between head channels in videotape recording and playback.

Saw Filter See "Surface Acoustic Wave Filter."

SBE See "Society of Broadcast Engineers."

SBS See "Satellite Business Systems."

SCA See "Subsidiary Communications Authority."

SCA Channel Program material of any kind which modulates a subcarrier well above the audio range, applied to a regular FM broadcast station or CATV system FM modulator.

SCA Modulator A relatively low power radio frequency transmitter which is modulated by an audio signal. The device that generates an SCA signal for transmission.

SCADA Acronym for security, CATV, and data.

Scalloping Horizontal displacement of lines in bands of approximately 16 per field resulting in a repetitive curving effect which may be apparent on vertical picture detail of a television picture from a playback of a video tape recording.

Scan The process of deflecting the electron beam.

Scanning The process of breaking down an image into a series of elements or groups of elements representing light values and transmitting this information in time sequence.

Scanning Line A single continuous horizontal narrow strip of picture area containing highlights, shadows and halftones. The process of scanning converts this to an electrical signal for transmission.

Scope See "Oscilloscope."

Scramble To interfere with an electronic signal or to rearrange its various component parts. In pay television, for example, the signal might be scrambled, and a decoder, also called a descrambler, might be necessary for the signal to be unscrambled so that only authorized subscribers would receive the clear signal.

Scrambler A device that transposes or inverts signals or otherwise encodes a message at the transmitter to make it unintelligible to a receiver not equipped with an appropriate descrambling device. Synonymous with encoder.

Scrape A continuous sound composed of a rapid series of clicks.

Scratch (Film) A break in the surface of the emulsion or the base material of the film.

Screw Anchor A type of anchor for utility pole guys that is screwed into the ground during its installation.

Scrolling A property of most alphanumeric video display terminals. If the screen of such a video terminal is filled, it will move the entire display image upward, either at a smooth pace or one line at a time, so that room is continuously made at the bottom of the screen for new information.

SCTE See "Society of Cable Television Engineers."

SECAM A color television system developed and used in France and adopted by a number of countries for use as their national standard.

Second Audio Program (SAP) In a BTSC-encoded television sound carrier, a monaural audio subcarrier that can be used to transmit supplemental foreign language translation audio or other information.

Second Harmonic In a complex wave, a signal component whose frequency is twice the fundamental, or original, frequency.

Second Order Beat Even order distortion product created by two signals mixing or beating against each other.

Secondary Service Area Broadcast area served by the skywave and not subject to objectionable interference. Subject to intermittent variations in intensity, usually applies to AM broadcasting.

Secondary Station Any station except a Class 1 station operating on a clear channel.

Security Systems A service provided by some communications companies to interconnect homes to fire departments, police, or intermediate agencies.

Semiconductor A material whose resistivity lies between that of conductors and insulators, e.g., germanium and silicon. Solid state devices such as transistors, diodes, photocells, and integrated circuits are manufactured from semiconductor materials.

Semiconductor Memory Computer memory using solid state devices instead of mechanical, magnetic, or optical devices.

Send Reference Station—Television The transmit earth station of a multiple destination satellite television transmission.

Sensor A device that converts measurable elements of a physical process into data meaningful to a computer.

Sensor-Based Pertaining to the use of sensing devices, such as transducers or sensors, to monitor a physical process.

Serial Input/Output Data transmission in which the bits are sent one by one over a single wire.

Serial Transmission In data communication, transmission at successive intervals of signal elements constituting the same telegraph or data signal. The sequential elements may be transmitted with or without interruption, provided that they are not transmitted simultaneously; for example, telegraph transmission by a time-divided channel.

Series A group of two or more programs which are centered around, and dominated by, the same individual, or which have the same, or substantially the same, cast of principal characters or a continuous theme or plot.

Serrated Pulses A series of equally spaced pulses within a pulse signal. For example, the vertical sync pulse is serrated in order to keep the horizontal sweep circuits in step during the vertical sync pulse interval.

Serration (Distortion) A picture condition in which vertical or nearly vertical lines have a ragged appearance.

Service Bureau An organization that packages its services so that all users have to do is to supply the input data and pay for the results.

Service Circuit See "Engineering Service Circuit."

Servomechanism An automatic device that uses feedback to govern the physical position of an element.

Setup The separation in level between blanking and reference black levels.

Setup Interval The interval between the blackest portions of the video waveform and the blanking wave-form in a video camera.

SGDF See "Supergroup Distribution Frame."

Shading Spurious variations in the tonal gradient of a television picture.

Share The measurement of television program audience that represents a percentage of televisions in use tuned to a particular program.

Sheath The outer conductor, or shield, of coaxial cable.

Sheath Current Unwanted electrical energy traveling along the strand and outer surface of the shield of coaxial cable.

Shield The outer conductor of coaxial cable, which is separated from the center (inner) conductor by a dielectrical material.

Shielding (1) In computer graphics, blanking of all portions of display elements falling within some specified region; (2) In CATV, shielding refers to the coaxial cable outer conductor, or the effectiveness of that conductor as an electromagnetic barrier.

SHL See "Studio-to-Headend Communications Link."

Shop-at-Home Channels Cable service that gives viewers the opportunity to view and/or purchase products.

Short-Time Distortion The linear waveform distortion of time components from 0.1265 to 1.09 microseconds.

Shrink Tubing A plastic-based tubing which, when heated to a critical temperature, will shrink and form a weatherproof seal. Generally, heat shrink tubing is applied to CATV connectors to protect the connection from any possibility of water infusion.

Shutter Bar One thin, light-toned horizontal line or two thin, light-toned horizontal lines about one-half the picture height apart which move slowly up or down in a television picture. Synonymous with hum bars.

Sidebands Additional frequencies generated by the modulation process, which are related to the modulating signal and contain the modulating intelligence.

Sign-On Procedure The process of connecting with a remote computer, including the provision of identification details and security access.

Signal Generator An electronic instrument which produces audio or radio frequency signals for test, measurement or alignment purposes.

Signal Leakage See "Leakage."

Signal Level Amplitude of signal voltage measured across 75 ohms, usually expressed in decibel millivolts.

Signal Level Meter (SLM) See "Field Strength Meter."

Signal-to-Noise Ratio The ratio, expressed in decibels, of the peak voltage of the signal of interest to the root-mean-square voltage of the noise in that signal.

Significantly Viewed Signal Signals that are significantly viewed in a county (and thus are deemed to be significantly viewed within all communities of the county) are those viewed in other cable television households as follows: (1) for full or partial networks station—a sharing of viewing hours of at least 3 percent (total week hours), and a net weekly circulation of at least 25 percent; and (2) for an independent station—a share of viewing hours of at least 2 percent (total weekly hours), and a net weekly circulation of at least 5 percent.

Silicon Chip A wafer of silicon providing a semiconductor base for a number of electrical circuits.

Simplex A circuit capable of transmission in one direction only. Contrast with "Half Duplex" and "Full Duplex."

Simulsweep An electronic instrument used to measure the broadband frequency response of a cable system. Synonymous with summation sweep.

Sine Square Pulse A test signal used to

evaluate short-time wave-form distortions.

Sing Any high-pitched spurious audible tone or a high-frequency spurious audio signal.

Single Channel Amplifier A narrowband amplifier which is tuned to boost the signal strength of one particular television channel.

Single Channel Antenna An antenna whose elements are cut to a precise length so as to be resonant at the desired frequency in order to handle one channel very well but be very inefficient at other channels.

Single-Sideband Transmission That type of carrier transmission in which one sideband is transmitted, and the other is suppressed. The carrier wave may be either transmitted or suppressed.

64K RAM Random-access memory that stores more than 64,000 bits of computer data on a tiny slice of silicon, four times as much as the 16K RAM—the previous generation chip.

Skew The angular deviation of recorded binary characters from a line perpendicular to the reference edge of a data medium.

Skewing Horizontal displacement of video information in bands of approximately 16 lines per field producing a sawtooth effect which is most apparent on vertical picture detail of a television picture originating from the playback of a video tape recording.

Skip Field Recording A process applicable to helical recorders that have more than one video head. In the case of a two-headed recorder, every second television field is recognized. During playback, every recorded field is reproduced twice. This process is used to reduce recording tape consumption.

SLM Signal level meter. See "Field Strength Meter."

Slope The difference in gain of a net-

work between the ends of a band.

Smaller Television Market The specified zone of a commercial television station licensed to a community that is not listed as a top 100 market community in Federal Communications Commission regulations.

SMATV See "Satellite Master Antenna Television System."

Smearing Blurring of the vertical edges of images in a television picture.

SMPTE See "Society of Motion Picture and Television Engineers."

Snow See "Noise."

Sniffer Comsonics' trade name for a specific type of radio frequency signal leakage detection equipment.

Society of Broadcast Engineers (SBE) A professional society serving the professional interests of broadcast engineers. Formerly Institute of Broadcast Engineers.

Society of Cable Television Engineers (SCTE) A professional society, serving the technical needs of the cable television industry, whose goal is to elevate the technical competence of its members for their own benefit in personal career growth as well as for the benefit of the company that employs them.

Society of Motion Picture and Television Engineers (SMPTE) An organization concerned with the engineering aspects of motion pictures, television, instrumentation, high-speed photography and the allied arts and sciences.

Software The non-physical requirements of a system that can be measured in information terms. For example, a computer program, an audio or visual script or program, a code sequence or an address.

Solid State A class of electronic components utilizing the electronic or magnetic properties of semiconductors.

Sort (Sorting) To segregate items into groups utilizing the electric or magnetic properties of semiconductors.

Sound Program Local Channel A channel used to transmit the audio portion of a television program between two points within a given urban area.

The Source A computer network service.

Space Diversity See "Diversity Reception."

Spacing Length of cable between amplifiers usually expressed in equivalent decibels of gain required to overcome cable losses at the highest television channel or frequency carried in the system; for example, 22-dB spacing.

Span A term used to indicate distance between two supporting structures.

Span Clamp "O" A clamping device used to take off house drop wires or cable from the supporting strand.

Special Facility Facility ordered and supplied on a per-occasion basis.

Specialized Common Carrier (1) A company authorized by a government agency to provide a limited range of telecommunications services. Examples of specialized common carriers are the value-added networks; (2) Those common carriers not covered in the original federal communications legislation.

Specified Zone of a Television Broadcast Station The area extending 35 air miles from the reference point in the community to which that station is licensed.

Spectrum Analyzer A scanning receiver with a display that shows a plot of frequency versus amplitude of the signals being measured. Modern spectrum analyzers are often microprocessor controlled and feature powerful signal measurement capabilities.

Spin The rapid switching among premium channels by subscribers, who are, in effect, "spinning" the channels.

Splice A mechanical/electrical connection to join two wires or cables together.

Split Screen The division into sections of a display surface in a manner that allows two or more programs to use the display surface concurrently.

Splitter Usually a hybrid device, consisting of a radio frequency transformer, capacitors and resistors, that divides the signal power from an input cable equally between two or more output cables. See also "Power Splitter."

Sports Blackout The local and/or regional non-televising, as required by federal regulation, of sporting events which have not been sold out. The purpose of a sports blackout is to protect ticket sales for the event.

Sprocket Hole Noise A repetitive noise occuring at the frequency of the film sprocket perforations.

Spurious Signals Any undesired signals such as images, harmonics, and beats.

SRL See "Structural Return Loss."

Stacked Antenna Array A group of identical antennas physically grouped and connected electrically for greater gain and directivity.

Staircase Video Wave-Form A test signal consisting of a series of discrete steps of picture level resembling a staircase.

Standard Broadcast Band The band of frequencies extending from 535 to 1605 KHz, usually called AM.

Standard Broadcast Channel The band of frequencies occupied by the carrier and two side bands of a broadcast signal with the carrier frequency at the center, usually 10 KHz wide.

Standard Broadcast Station A broadcasting station licensed for the transmission of radio telephone emissions primarily intended to be received by the general public and operated on a channel in the band 535-1605 KHz.

Standard Television Signal A signal which conforms to the television transmission standards of the Federal Communications Commission. See also "NTSC Video Signal."

Standby Facilities Facilities furnished for use as replacement in the event of failure or faulty operation of normally used facilities.

Standby Generator A fuel-powered (e.g., gasoline, propane, diesel) generator used to backup electrical power in the event of utility power failure.

Standby Power Supply A stepdown alternating current (AC) transformer which converts 120 volts AC to a lower AC voltage (30 or 60 volts) to be carried on the coaxial cable along with the cable signals to power active devices in the distribution plant. In addition, batteries and an inverter are included to provide backup power in the event of utility power (120 VAC) failure.

Star Network A network configuration in which there is only one path between a central or controlling node and each end-point node.

Start Bit A bit used to signal the arrival of other characters in asynchronous data transmission.

State Association Trade associations serving the CATV industry at the state level. Their goals are to represent their members in lobbying efforts with the state legislatures, to raise the level of awareness of cable television on a statewide basis, and to provide a forum for discussion of issues that affect their members. Membership includes cable operators, suppliers to the CATV industry, and programmers within each state.

Step-by-Step System A type of line-switching system that uses step-by-step switches.

Stereo TV Sound See "BTSC."

Stereophonic Giving, relating to, or constituting a three-dimensional effect of auditory perspective, by means of two or more separate signal paths in FM and AM broadcasting.

Stereophonic Separation The ratio, expressed in decibels, of the electrical signal caused in the right (or left) stereophonic channel to the electrical signal caused in the left (or right) stereophonic channel by the transmission of only a right (or left) signal.

Stereophonic Subchannel The band of frequencies from 23 to 53 KHz containing the stereophonic subcarrier and its associated sidebands.

Still Video In telecommunications, a technique whereby the telephone is linked to a screen and calls are accompanied by or interspersed with static images, permitting a lower bit rate than is needed for making pictures and/or higher resolution.

STL See "Studio-to-Transmitter Communications Link."

Store and Forward Mode (1) A manner of operating a data network in which packets or messages are stored before transmission toward the ultimate destination; (2) In CATV, this technique is used in pay-per-view program service in the subscriber terminal device until it is retrieved later by the addressing computer for subsequent billing purposes.

Strand A steel support wire to which the coaxial cable is lashed in aerial installations. Synonymous with suspension strand.

Stratosphere The layer of atmosphere directly above troposphere; it extends to a height of about 40 miles.

Streaking A picture condition in which objects appear smeared or extended horizontally beyond their normal boundaries.

Street Side The vehicle access side of a pole to which cables are attached.

Strip Amplifier A single television channel amplifier, which has relatively flat frequency response and uniform gain throughout the channel passband. Strip amplifiers are sometimes used in small systems as a low-cost substitute for a headend processor.

Structural Return Loss (SRL) Return loss characteristics of coaxial cable that are related to periodic discontinuities within the cable itself.

Studio A specially designed room with associated control and monitoring facilities used by a broadcaster for the origination of radio or television programs.

Studio-to-Headend Link (SHL) A coaxial or radio link that connects a radio or television studio directly to a cable television system headend.

Studio Transmitter Link (STL) A radio link that connects a radio or television studio to a remote transmitter location.

STV See "Subscription Television."

Sub-Band See "Sub-VHF Channels."

Subcarrier A carrier used to modulate information upon another carrier; for example, the difference channel subcarrier in an FM stereo transmission.

Sub-Control ITC The international television center (ITC) at the originating end of an international television transmission.

Sub-Control Station A station at the transmitting end of a sound program or television circuit section, link or connection.

Subscriber A customer who pays a fee for cable television service.

Subscriber Drop A cable which connects the tap or coupler of a feeder cable to a subscriber's premises and terminal, often a television set.

Subscriber Tap See "Subscriber Terminal."

Subscriber Terminal The cable television system terminal to which a subscriber's equipment is connected. Synonymous with subscriber tap.

Subscription Television (STV) The broadcast version of pay television. Not a cable service, it is distributed as an over-the-air broadcast signal. Its signals are scrambled and can be decoded only by a special device attached to the television set for a fee. STV contains no commercials.

Subsidiary Communications Authority (SCA) Multiplexed signals on the regular FM broadcast band.

Subsplit A cable-based communications system that enables signals to travel in two directions, forward and reverse, simultaneously with upstream (reverse) transmission from 5 to about 30 MHz, and downstream (forward) transmission above about 50 MHz. Exact crossover frequencies vary from manufacturer to manufacturer.

Sub-VHF Channels Television channels, usually between 5.75 and 47.75 MHz, or at frequencies lower than Channel 2. Synonymous with sub-band.

Subvideo-Grade Channel A channel of bandwidth narrower than that of video-grade channels.

Suckout (1) The result of the coaxial cable's center conductor, and sometimes the entire cable, being pulled out of a connector because of contraction of the cable; (2) A sharp reduction of the amplitude of a relatively narrow group of frequencies within the cable system's overall frequency response.

Summation Sweep See "Simulsweep."

Sunrise and Sunset For each particular location and during any particular month, the time of sunrise and sunset as specified in the instrument of authorization; usually applies to AM broadcasting.

Super Blanking Pulse A special blanking

signal generated in kine recording equipment and used to override and remove a part of the picture normally presented by a picture monitor.

Superband The band of cable television channels J through W lying between 216 and 300 MHz.

Supercomputers The fastest and most powerful computing systems that are available at any given time.

Supergroup Distribution Frame (SGDF) In telephone frequency division multiplexing, the SGDF provides terminating and interconnecting facilities from group modulator output, group demodulator input, supergroup modulator input, and supergroup demodulator output circuits of the basic supergroup spectrum of 315 to 552 KHz.

Superheterodyne Reception A method of receiving radio waves that operates on the principle of mixing the received signal with a local oscillator signal to produce an intermediate frequency prior to detection.

Superstation A broadcast station whose signal is transmitted via satellite or terrestrial microwave to cable systems, which use the advertiser-supported programming as part of their basic service. Examples of superstations are WTBS (Atlanta), WWOR (New York), and WGN-TV (Chicago).

Supertrunk A signal transportation cable that normally is not used to provide service directly to subscribers, but rather to link a remote antenna site with a headend, to link a headend with the distribution system, or to interconnect hub sites.

Supervisor In computers, the monitor or master control program that controls the very low level operations of a computer system.

Supervisor Call Instruction An instruction that interrupts the program being executed and passes control to the supervisor so that it can perform a specific service indicated by the instruction.

Surface Acoustic Wave Filter Electronic filters that generally provide performance superior to that of conventional passive filters. Term is frequently shortened to "saw filter."

Suspension Strand See "Strand."

Sweep Generator An electronic instrument whose output signal varies in frequency between two preset or adjustable limits, at a rate that is also adjustable. This "swept" signal is used to perform frequency response measurements when used in conjunction with appropriate peripheral accessories.

Switched Circuit A circuit that may be temporarily established at the request of one or more stations.

Switched Network Any network in which switching is present and is used to direct messages from the sender to the ultimate recipient. Usually switching is accomplished by disconnecting and reconnecting lines in different configurations in order to set up a continuous pathway between the sender and the recipient.

Switched System A communications system (such as a telephone system) in which arbitrary pairs or sets of terminals can be connected together by means of switched communications lines.

Switching Center A location that terminates multiple circuits, and is capable of interconnecting circuits or transferring traffic between circuits.

Symmetrical Channel A channel pair in which the sending and receiving directions of transmission have the same data signalling rate.

Sync An abbreviation for the words "synchronization," "synchronizing," etc. Applies to the synchronization signals, or timing pulses, which lock the electron beam of a picture monitor in step, both horizontally and vertically, with the electron beam of the pickup tube. The color sync signal (NTSC) is also known as the color burst.

Sync Compression A reduction in the amplitude of the sync signal, with respect to the picture signal.

Sync Generator An electronic device that supplies common synchronizing signals to a system of several video cameras, switchers, and other video production equipment, ensuring that all will be "locked" to a master timing reference.

Sync Level The level of the tips of the synchronizing pulses, usually 40 IRE units from blanking to sync tip.

Synchronous For transmission, operation of the sending and receiving instruments continuously at the same frequency.

Synchronous Data Network A data network that uses a method of synchronization between data circuit-terminating equipment (DCE) and the data-switching exchange (DSE), and between DSEs, the data-signaling rates being controlled by timing equipment.

Synchronous Idle Character A transmission control character used by synchronous data-transmission systems to provide a signal from which synchronism or synchronous correction may be achieved between data terminal equipment, particularly when no other character is being transmitted.

Synchronization The maintenance of one operation in frequency and/or in phase with another.

Synchronization Pulses Specific pulses used to cause the electron beam of a television picture tube to operate in synchronism with the electron beam of the scanning device of the program source equipment.

Syndicated Exclusivity An obsolete Federal Communications Commission rule that regulated the cable system's televising of syndicated programs that were also being broadcast by local television stations who had the programs under exclusive contract.

Syndicated Program Any program sold, licensed, distributed, or offered to television station licensees in more than one market within the United States for non-interconnected television broadcast exhibition.

Syntax Error A mistake in the formulation of an instruction to a computer.

SYSRES The disk pack which contains the supervisor.

System Impedance The resistance and reactance opposing the current flow in the system. For CATV the impedance is 75 ohms. See also "Impedance."

System Noise That combination of undesired and fluctuating disturbances within a cable television channel that degrades the transmission of the desired signal.

System Operator The individual, organization, company, or other entity that operates a cable television system.

T

Tagging The transmission of special coded signals within a television channel to identify a particular program.

TAMI See "Television Accessory Manufacturers Institute."

Tandem Data Circuit A data circuit that contains more than two data circuit-terminating equipment in series.

Tandem System A system network where data proceeds through one central processor into another. This is the system of multiplexers and master/slave arrangements.

Tandem Trunk A trunk extending from a telephone central office (or tandem office) to a tandem office and used as part of a connection between telephone stations in different central offices.

Tap See "Multitap."

Tape Drive A mechanism for controlling the movement of magnetic tape. This mechanism is commonly used to move magnetic tape past a read head or write head, or to allow automatic rewinding.

Tape Dump The transfer of complete contents of information recorded on tape to a computer or another storage medium.

Tariff The published rate for a specific unit of equipment, facility, or type of service provided by a telecommunication facility. Also, the vehicle by which the regulating agencies approve or disapprove such facilities or services. Thus, the tariff becomes a contract between the customer and the telecommunications facility.

TASO See "Television Allocation Study Organization."

TCC See "Telephone Coordinating Circuit."

TDR See "Time Domain Reflectometer."

Tearing One or more horizontal lines, in a television picture, horizontally displaced in an irregular manner.

Telco Generic name for telephone companies.

Telecommunications Transmission and reception of signals by electromagnetic means.

Telecoms Originally a British term, a telephone installation for a partnership or small business in which any extension can make or receive calls with intercom facilities between extensions.

Teleconference A meeting between people who are remote from each other but are linked together by a two-way video and audio telecommunications system.

Telecopier A unit for facsimile transmission.

Teledata A unit that introduces parity bits to punched paper tape for transmission. The receiving unit checks parity for code accuracy and repunches paper tape with valid data.

Telefax Linking photocopying units for the transmission of images.

Telegraph A system employing the interruption or change in polarity of direct current for the transmission of signals.

Teleinformatics Data transfer via telecommunication systems.

Telemanagement A service featuring computerized management of a customer's long-distance system, automatically routing each call over the least costly line available at the time the call is made and logging the call for accounting control.

Telemeter To transmit digital or analog metering data by telecommunication facilities. For example, data can be telemetered from a missile and recorded at a ground station.

Telemetry Sensing or metering of remote operating systems by a receiving unit that converts transmitted electrical signals into units of data.

Telenet U.S.-based packet-switching service (GTE).

Telepak A leased channel offering of telephone companies and Western Union providing specific-sized bundles of voice-grade, telegraph-grade, subvoice-grade, and broadband channels between two locations. Mileage charges are constant for each mile rather than regressive as in conventional single-leased lines.

Telephone Coincidental Survey A method of measuring television program audience size using random telephone calls to television households.

Telephone Coordinating Circuit (TCC) A circuit used for point-to-point speech communications between or among members of a broadcaster's staff.

Telephone Coupler A device for putting a regular telephone handset into service as a modem. Usually it works acoustically, but it may also work inductively.

Telephone Data Set A device connecting a telephone circuit to a data terminal.

Telephone Dialer Under program control, a circuit that divides the output of an on-chip crystal oscillator, providing the tone frequency pairs required by a telephone system. The tone pairs are chosen through a latch by means of a binary-code-decimal (BCD) code from the bus.

Telephone Set The terminal equipment on the customer's premises for voice telephone service. Includes transmitter, receiver, switch hook, dial, ringer, and associated circuits.

Telephony The use or operation of an apparatus for transmission of sounds between widely removed points with or without connecting wires.

Teleport (1) A project developed by the Port Authority of New York and New Jersey, Merrill Lynch & Company and the Western Union Corporation to provide the New York City metropolitan area with satellite communications; (2) A generic term referring to a facility capable of transmitting and receiving satellite signals for other users.

Teleprocessing See "Remote-Access Data Processing."

Teleprocessing Terminal Terminal used for on-line data transmission between remote process locations and a central computer system. Connection to the computer system is achieved by a data-adaptor device or a transmission control.

Telescreen A two-way audiovisual television used to monitor and control remote activities.

Teleshopping Remote selection and purchase of goods and services via electronic channels (e.g., telephone, CATV, viewdata).

Teletext Generic term for one-way information retrieval systems that employ a broadcast signal to carry digitally encoded textual and graphic information that is constantly transmitted in a continuous cycle. Users of a teletext service "grab" pages from the transmission cycle using a keypad similar to that used in videotex systems.

Teletext Service Transmission of information over communications networks between typewriter-like terminals.

Teletype The trademark of Teletype Corp., usually referring to a series of different types of teleprinter equipment (such as tape punches, reperforators, and page printers) used for telecommunications.

Television The electronic transmission and presentation of pictures and sounds.

Television Accessory Manufacturers Institute (TAMI) A trade association, now defunct.

Television Allocation Study Organization (TASO) The organization that created television picture impairment standards that assign numerical grades to the subjective picture quality of a television signal. TASO Grade 1 (excellent) indicates the picture is of extremely high quality, with no perceptible interference. TASO Grade 2 (fine) indicates the picture is of a high enough quality to provide enjoyable viewing, with impairment just perceptible. TASO Grade 3 (passable) indicates the picture is of acceptable quality; impair-

ment is definitely perceptible but not objectionable. TASO Grade 4 (marginal) indicates a picture of poor quality, with somewhat objectionable impairment.

Television Authority—Receive See "Broadcasting Authority—Receive."

Television Authority—Send See "Broadcasting Authority—Send."

Television Broadcast Band The frequencies in the band extending from 54 to 890 MHz assignable to television broadcast stations. Channels 2-13 usually operate in the 54 to 216 MHz range.

Television Broadcast Booster Station A station in the broadcasting service operated for the sole purpose of retransmitting the signals of a television broadcast station by amplifying and reradiating signals which have been received directly through space, without significantly changing any characteristics of the incoming signal other than its amplitude; usually applied to television broadcasting station service, operates on the same channel as the originating station.

Television Broadcast Translator Station A station in the broadcast service operated for the purpose of retransmitting the signals of a television broadcast station, another television broadcast translator station, or a television relay station, by means of direct frequency conversion and amplifications of the incoming signals without significantly altering any characteristic of the incoming signal other than its frequency and amplitude. A so-called "TV booster" using UHF and VHF channels.

Television Channel The range or band of the radio frequency spectrum assigned to a television station. In Canada and the United States, the standard bandwidth is 6 MHz.

Television Demodulator A television

receiver which does not have a picture tube and its associated circuits, used to derive a video and audio signal from a television channel.

Television Household A household having one or more television sets.

Television Intercity Relay Station A fixed station used for intercity transmission of television program material and related communications for use by television broadcast stations.

Television Market The community or group of communities served by commercial television broadcast signals from one or more television stations located within that area.

Television Modulator A low-power television transmitter used for transmitting locally originated programs, closed circuit television signals, video-taped programs, and demodulated off-the-air television programs.

Television Operating Center (TOC) A communications company location where television signals are switched and monitored.

Television Penetration The percent of total homes having one or more television sets.

Television Pickup Station A mobile land station that passes television program material and related communication transmissions from remote "scene of events" television broadcast studios to television broadcast stations.

Television Receive-Only Earth Station (TVRO) The receiving antenna dish, or the complete package of dish and receiver.

Television Service Point The nearest operation point in a broadcaster's or communications company's facilities to the interconnection point.

Television Sound Modulator An electronic device used to transmit music and other audio programs for reception by sub-

scribers with television receivers, usually in the FM band of frequencies. See also "FM Modulator."

Television STL Station A fixed station used for the transmission of television program material and related communications from the studio to the transmitter of a television broadcast station.

Television Translator A radio repeater station which intercepts television signals and rebroadcasts them on a locally unoccupied channel.

Television Translator Relay Station A fixed station used for relaying signals of television broadcast stations to television broadcast translator stations.

Television Transmission Standard The standards which determine the characteristics of a television signal as radiated by a television broadcast station, according to Federal Communications Commission rules.

Television Transmitter The radio transmitter or transmitters for the transmission of both visual and aural signals.

Telidon The Canadian Department of Communications system for character coding and display in videotex systems.

Telpak An AT&T corporate name designating its service for leasing wide-band channels.

Terminal (1) Generally, connection point of equipment, power or signal; (2) Any "terminating" piece of equipment such as a computer terminal.

Terminal Controller A hard-wired or intelligent (programmable) device which provides detailed control for one or more terminal devices.

Terminal Isolation The attenuation, at any subscriber terminal, between that terminal and any other subscriber terminal in the cable television system.

Terminator A resistive load for an open coaxial line used to eliminate reflections and to terminate a line in its characteristic impedance.

Thermal Equalizer (1) A network of temperature-sensitive components which cause a loss inverse to the losses suffered in the cable caused by changes in temperature; (2) A frequency equalizer controlled by pilot channels.

Third Harmonic In a complex wave, a signal component whose frequency is three times the fundamental, or original, frequency.

Third Order Beat See "Triple Beat."

Tier or Tiered Service Different packages of programs and services on cable television systems, for different prices; a marketing approach that divides services into more levels than simply "basic" service and "pay" services. See also "Pay Cable" and "Basic Cable Service."

Tilt The difference in cable attenuation or amplifier gain between lower and higher frequencies on the cable system. Synonymous with output tilt.

Tilt Compensation The action of adjusting, manually or automatically, amplifier frequency/gain response to compensate for different cable length frequency/attenuation characteristics.

Time Base Corrector A device used to correct timing or synchronizing errors caused by tape stretch and other problems during video tape playback.

Time-Division Multiplexing See "Multiplexing."

Time Domain Reflectometer A test device used to test the condition of coaxial cable. The device launches a pulse signal into the cable under test and displays pulse reflections from the cable on a CRT display tube similar to radar. The amplitude of a reflected pulse is proportional to the degree of electrical imperfection, and the reflected pulse's position on the horizontal time line is proportional to the distance from the launch point where the imperfection exists.

Time Lock Promotions Marketing campaigns that take place during a predetermined time period with a specific beginning and end.

Time Sharing Pertaining to the interleaved use of time on a computer system that enables two or more users to execute computer programs concurrently.

TOC See "Television Operating Center."

Toegepaste Viewdata System (TVS) A viewdata system in Holland operated by VNU, a major publishing company.

Tone Switch A solid state switching device that responds to audio cue tones added to satellite programs for the purpose of operating commercial insertion and channel timeshare switching equipment.

Top 100 Markets The 100 largest television markets in the United States, as defined by the television industry.

Total Households The sum of all living units (occupied or unoccupied) in the area of concern. A living unit includes a single dwelling unit (house), individual apartments, or any group of one or more rooms used as a domicile.

Touchtone Developed by AT&T, the replacement of the conventional telephone dial with a panel of buttons, which, when pushed, generate tones which operate switching devices at the telephone exchange. These tones can be utilized to yield input to a computer.

Trace The CRT display produced by the moving electron beam. Usually used in context of test equipment but acceptable also in context of television scan lines.

Traffic Cone See "Safety Cone."

Traffic Loading Patterns The statistically definable load placed on a facility such as a telephone switching exchange. A multipoint-to-point traffic load, such as that created by a radio call-in program, may cause more demand for switches than there are switches available.

Trailing Blacks See "Following Blacks."

Trailing Whites See "Following Whites."

Transaction (1) An operational unit of processing at the application level; a complete step of data processing; (2) A logical grouping of messages in both directions between originating and target hosts in a network; (3) A two-way communication in teleprocessing enabling the user to interact with a database.

Transaction File A file containing relatively transient data that, for a given application, is processed together with the appropriate master file.

Transceiver (1) A terminal that can both transmit and receive data; (2) A radio that can both transmit and receive.

Transducer A device that is actuated by energy from one system and supplies energy in any other form to a second system, such as a microphone changing sound energy to electrical energy or a television camera changing light images to electrical signals.

Transients Usually random electrical phenomenon, such as current or voltage pulses of brief duration. Transients often are fast risetime, high amplitude impulses that can damage electronic equipment.

Transistor An active semiconductor device capable of amplification, oscillation, and switching action. The transistor, commonly a three-terminal component, has replaced the vacuum tube in most applications. The name is a combined word derived from "transfer resistor." See also "Semiconductor."

Transit County A county through which an international television link is routed without the use of program

material by that county.

Translator (1) A piece of processing equipment housed in the system's headend responsible for the reception and retransmission of data signals; (2) A station in the broadcasting service operated for the sole purpose of retransmitting the signals of a television station by amplifying and re-radiating those signals which have been received directly through space, without signficantly altering any characteristic of the incoming signal other than its amplitude and frequency.

Transmission The sending of information (signals) from one point to another.

Transmission Lines The electrical conductors, normally coaxial cables, in CATV systems, for transporting radio frequency signals and in some cases electrical power.

Transmultiplexer Equipment that transforms signals derived from frequency-division multiplex equipment (such as group or supergroup) to time-division multiplexed signals having the same structure as those derived from pulse code modulation multiplex equipment (such as primary or secondary pulse code modulation multiplex signals) and vice versa.

Transparent (1) In data transmission, pertains to information that is not recognized by the receiving program or device as transmission control characters; (2) In communications, a circuit or device is transparent to the signal when that signal is transmitted with little or no distortion to its original form.

Transparent Transmission (1) Transmission where the transmission medium does not recognize control characters or initiate any control activity; (2) Transmission where the baseband signal remains essentially unaltered or undistorted through the transmission medium.

Transponder That portion of a satellite used for reception and retransmission of a signal or signals.

Trap (1) A passive device used to block a channel or channels from being received by a CATV subscriber (negative trap), or used to remove an interfering carrier from a channel that a subscriber wants to receive (positive trap); (2) An unprogrammed, conditional jump to a specified address that is automatically activated by hardware, a recording being made of the location from which the jump occurred.

Trapping (1) The installation of negative traps in a CATV system to prevent unauthorized reception of one or more signals; (2) A unique feature of some computers enabling an unscheduled jump (transfer) to be made to a predetermined location in response to a machine condition.

Traveling-Wave Tube An efficient, high-power radio frequency amplifying device used in microwave radio transmitters.

Tree Network A design for a cable system in which signals are disseminated from a central source. The configuration resembles that of a tree, in which the product from the root (headend) is carried through the trunk and then through the branches (feeders) to the individual stems (drops), which feed each individual leaf (terminal). See also "Hub Network."

Triple Beat Odd order distortion products created by three signals, mixing or beating against each other, whose frequencies fall at the algebraic sums of the original frequencies. Synonymous with third order beat.

Troposphere The portion of the earth's atmosphere extending from sea level to a height of about six miles; the weather layer.

Trophospheric Scatter—Communications A term applied to over-the-horizon radio communication systems making use of the random irregularities of the dielectric constant of the atmosphere which can reflect or bend radio signal paths in such a manner as to achieve considerably longer distances than normal.

Troubleshoot To detect, locate, and eliminate errors in computer programs or faults in hardware or electrical circuits.

Trucking Moving a television camera dolly from one side of a scene to another laterally. Synonymous with dollying.

Truncate To terminate a computational process in accordance with some rule; for example, to end the evaluation of a power series at a specified term.

Truncation The deletion or omission of a leading or trailing portion of a string in accordance with specified criteria. The termination of a computer process before its final conclusion or natural termination, if any, in accordance with specified rules.

Trunk The main distribution lines leading from the headend of the cable television system to the various areas where feeder lines are attached to distribute signals to subscribers.

Trunk Amplifier An amplifier inserted into a trunk line. A weak input signal is amplified before being retransmitted to an output line, usually carrying a number of video voice or data channels simultaneously. Amplifiers increase the range of a system. Usually, trunk amplifiers must be inserted approximately every 1500 to 2000 feet.

Trunk and Feeder System See "Distribution System."

Trunk Line The main highway coaxial line of a CATV system which feeds signals from the headend to the community being served. Trunk lines are usually three-quarters to over one inch in diameter. Subscribers are never directly connected to trunk lines.

Tuner A device, circuit, or portion of a circuit that is used to select one signal from a number of signals in a given frequency range.

Turn-Key An arrangement where all aspects of building or rebuilding a cable system are handled by a single outside entity. Accordingly, all design, construction, splicing, alignment and testing is under management responsibility of the entity until the total job or phase is complete and turned over to the cable system management.

TVRO See "Television Receive-Only Earth Station."

TVS See "Toegepaste Viewdata System."

Twinlead Cable A cable composed of two insulated conductors laid parallel and either attached to each other by the insulation or bound together with a common covering.

Twisted Pair A pair of wires used in transmission circuits and twisted about one another to minimize coupling with other circuits. Paired telephone cable is made up of a few to several thousand twisted pairs.

Two-Way Capacity The bandwidth available for upstream or two-way communication.

Two-Way CATV A cable system capable of transmitting signals simultaneously in two directions, upstream and downstream, in either a subsplit, midsplit or high split frequency configuration. Two-way cable systems are not necessarily interactive cable systems.

Two-Way Videotex See "Viewdata."

TWX The name given by Western Union to its teleprinter exchange service, which provides real time direct connection between subscribers. TWX service is confined to North America, in contrast to Telex service, which is worldwide.

Tymnet U.S.-based packet-switching network.

U-V

UAR See "Universal Asynchronous Receiver."

UHF See "Ultra High Frequency."

UHF to VHF Converter An electronic device for receiving UHF signals and translating them to VHF signals for transmission into cable network.

UHF Translator (Signal Booster) A station in the broadcasting service operated for the sole purpose of retransmitting the signals of a UHF translator station by amplifying and re-radiating such signals which have been received directly through space, without significantly altering any characteristic of the incoming signal other than its amplitude and frequency. Usually translators are used to fill or extend the coverage area of a television station.

Ultra High Frequency (UHF) Corresponding to electromagnetic signals in the range from 300-3000 MHz; channels 14-83 on the television dial.

Ultraviolet Erasing Erasable programmable read-only memory (EPROM) chips erased by exposure to high-intensity shortwave ultraviolet light.

Unattended Operation Operation of a station by automatic means whereby the transmitter is turned on and off and performs its functions without on-site attention by a qualified operator, and is monitored remotely.

Unbalanced Channel An audio channel whose terminals are at different electrical potentials with respect to ground. Generally, one terminal is grounded to the equipment chassis and the other terminal potential rises and falls according to the audio channel voltage.

Unblanking Turning on the cathode-ray beam. See also "Blanking."

Underground Cable System A cable system in which the cable and associated equipment is below ground level.

Underground Compacting Auger A device for boring under streets, highways, sidewalks and railroads.

Underground Housing An environmental protection device used to house subscriber isolation units, passive distribution amplifiers in underground CATV systems. Synonymous with pedestal housing.

Undershoot An insufficient response to a unidirectional voltage change.

Universal Asynchronous Receiver (UAR) A circuit that converts parallel data to serial data and vice versa.

Upconverter See "Output Converter."

Upgrade (1) The addition, by the subscriber, of a premium program service or any other added service or product to the existing level of CATV service; (2) A major cable system improvement that usually includes replacing active components, and in some cases passive components, to expand the system's channel capacity. An upgrade is not generally as involved as a rebuild.

Uplink Ground-to-satellite transmission.

Upstream In a cable system, the direction from the subscriber terminals to the headend. Compare with "Downstream."

User The individual or entity who actually uses a computer terminal to access databases, information services or computer time.

Vacant Line A horizontal line in the vertical blanking interval on which no information is present.

Validation The checking of data for correctness, or compliance with applicable standards, rules and conventions.

Value Added Network (VAN) A data network operated in the United States by a firm which obtains basic transmission facilities from the common carriers; for example, the Bell System adds "value" such as error detection and sharing and resells the service to users. Telenet and Tymnet are examples of VANs.

VAN See "Value Added Network."

Variable Field A field in a record the length of which is determined by the number of characters required to store the data in a given occurrence of that field. The length may vary from one occurrence of a variable field to the next.

Vault A protective enclosure that houses cable system components, active and passive, in underground installations. See also "Underground Housing."

VBI See "Vertical Blanking Interval."

VCR See "Videocassette Recorder."

Vectorscope An electronic instrument that produces a visual display of a video wave-form's phase and amplitude characteristics.

Velocity of Propagation Velocity of signal transmission along a coaxial cable relative to the speed of light in free space.

Vendor Supplier.

Venetian Blind Effect Television picture interference, made up of numerous horizontal lines, which somewhat resembles a venetian blind. This type of distortion is caused by co-channel interference.

Vertical Blanking Interval (VBI) The unused lines in each field of a television signal, seen as a thick band when the television picture rolls over usually at the beginning of each field, which instruct the television receiver to get ready for the reception of the next field. Some of these lines may be used for teletext and captioning, or may contain specialized test signals.

Vertical Edge Effect The narrow bands of color subcarrier which appear at the vertical edges of color transitions in the picture.

Vertical Interval Reference Test Signal (VIRS) Test signals transmitted in the vertical blanking interval for remote transmitter monitoring, for on-line testing, or as a reference for automatic receiver circuits. Synonymous with insertion test signal and vertical interval test signal (VITS).

Vertical Interval Test Signal (VITS) See "Vertical Interval Reference Test Signal."

Vertical Resolution The maximum number of black and white horizontal lines that the system can resolve.

Vertical Retrace The return of the electron beam from the bottom to the top of the raster after completion of each field.

Vertical Riser (1) Cable running vertically within the customer's building, serving as the link between the drop cables in the customer's premises and the main data trunk; (2) The vertical cable run where the aerial plant goes either underground or above ground.

Very High Frequency (VHF) Corresponding to electromagnetic signals in the range from 30 to 300 MHz.

Very Large Scale Integration Over 10,000 transistors per chip.

Vestigial Sideband AM Amplitude modulation in which the higher frequencies of the lower sideband are not transmitted. At lower baseband frequencies, the carrier envelope is the same as that for normal double sideband AM.

Vestigial Sideband Modulation A form of amplitude modulation, lying between double sideband and single sideband, in which one sideband and a small vestige of the other sideband are transmitted. Vestigial sideband modulation is attractive for television transmission because it uses less bandwidth than double sideband and preserves the wave-form of the signal.

Vestigial Sideband Transmission A system of transmission wherein one of the generated sidebands is partially attenuated at the transmitter and radiated only in part.

VHF See "Very High Frequency."

VHF to VHF Translator An electronic device for receiving on one VHF channel and transmitting on another VHF channel. See also "Translator."

VHF Translator A television broadcast translator station operating on a VHF television broadcast channel. See also "Translator."

VHS See "Video Home System."

Video A term pertaining to the bandwidth and spectrum of the signal which results from television scanning and which is used to reproduce a picture.

Video Bandwidth (1) The maximum rate at which dots of illumination are displayed on a screen; (2) The occupied bandwidth of a video signal. For NTSC, that bandwidth is 4.2 MHz.

Video Camera A camera which converts images to electrical signals for recording on magnetic tape or live transmission.

Video Conferencing Teleconferencing where participants see and hear others at remote locations.

Video Data Integrator A terminal device made up of a keyboard and separable associated display, providing a terminal facility for conventional communications lines.

Video Home System (VHS) A popular videocassette format developed for home use.

Video Local Channel A channel used to transmit the composite video signal portion of a television signal between two points within a given urban area.

Video Monitor A device that is functionally identical to a television set, except that it has no channel selector. It receives its picture signal from an external source such as a videocassette recorder, videodisc or viewdata computer. Synonymous with video picture monitor.

Video Network Channel Facilities, including channels, used to transmit the composite video signal portion of a television signal between broadcasters' premises in different urban areas.

Video Pair A transmission cable consisting of two twisted parallel, insulated conductors, with extra separation, surrounded by polyethylene or another high frequency insulating material and enclosed in a metal shield jacket.

Video Picture Monitor See "Video Monitor."

Video Response System (VRS) An experimental videotex system in Japan.

Video Tape Plastic tape with magnetic coating, used to record (and re-record) and play back video and audio signals.

Video Tape Recorder (VTR) Electromechanical device used to record television sound and picture on magnetic-coated tape for playback on a television receiver.

Video Tape Recording The retention in magnetic form on tape of composite video signals and audio signals.

Video Test Signal Generator An electronic device capable of generating a number of different video test patterns to facilitate testing and alignment of video equipment.

Video Transmission Engineering Advisory Committee (VITEAC) A U.S. committee, composed of the major television network broadcasters and the Bell Telephone System entities, which carries out long-range engineering coordination of the U.S. television network.

Video Wave-Form That part of the waveform which is produced by the camera tube and contains picture information only.

Videocassette Video tape in a container that provices easy loading, handling and unloading of this small self-contained reel-to-reel tape storage format.

Videocassette Recorder (VCR) An electronic device capable of playing or recording video tape in cassette format.

Videodisc A recordlike device storing a large amount of audio and visual information that can be linked to a com-

puter; one side can store the pictures and sounds for 54,000 separate television screens.

Videodisc Microprocessor A microprocessor that facilitates the interfacing of a videodisc player with computers and other data processors.

Videograph High-speed cathode-ray printer.

Videograph Display Computers that can draw pictures using dots or lines; computers that can rotate objects, showing them in perspective, moving them around, stretching or shrinking them.

Videotex The generic term used to refer to a two-way interactive system(s) for the delivery of computer-generated data into the home, usually using the television set as the display device. Some of the more often used specific terms are "viewdata" for telephone-based systems (narrowband interactive systems); "wideband broadcast" or "cabletext" for systems utilizing a full video channel for information transmission; and "wideband two-way teletext" for systems which could be implemented over two-way cable television systems. In addition to the systems mentioned here, hybrids and other transmission technologies, such as satellite, could be used for delivery of videotex services on a national scale.

Viewdata Generic term used primarily in the United States and Great Britain to describe two-way information retrieval systems based on mainframe computers accessed by dumb or intelligent terminals whose chief characteristic is ease of use. Originally designed to use the telephone network, viewdata in the United States is being implemented over other distribution media such as coaxial. Viewdata's salient characteristic is the formatting, storing, and accessing of screens (sometimes called frames or pages) of alphanumeric displays for retrieval by users according to a menu or through use of keyboard search. A two-way form of videotex.

Viewing Area The area of the phosphor screen of a CRT which can be excited to emit light by the electron beam.

Viewtron Experimental viewdata system established as a joint venture between AT&T and Knight-Ridder newspapers.

VIRS See "Vertical Interval Reference Test Signal."

Visual Carrier The picture portion of a television signal.

Visual Carrier Frequency The frequency of the carrier which is modulated by the picture information, which is 1.25 MHz above the bottom end of a television channel.

Visual Signal Level The peak voltage produced by the visual signal during the transmission of synchronizing pulses.

Visual Transmitter Radio equipment for the transmission of the visual signal only.

Visual Transmitter Power The peak visual power output when transmitting a standard television signal, which is controlled by the synchronizing pulse peak value of the video signal.

VITEAC See "Video Transmission Engineering Advisory Committee."

Voice-Grade Line A telephone line suitable for transmission of speech, digital or analog data, or facsimile, generally with a frequency range of 300 to 3000 cycles per second. Normal household telephone services use voice-grade lines.

Volatile Memory A storage medium in which information is destroyed when power is removed from the system.

Voltage Standing Wave Ratio (VSWR) The ratio of voltage peaks and minimums caused by the addition and subtraction of reflected signal wave present in a cable due to mismatch (faulty termination).

Volume Unit (VU) The logarithmic unit of measurement of fluctuating alternating current power, such as that of speech or music. Four milliwatts across a 600-ohm impedance corresponds to zero VU.

Von Newmann Sort In a sort program, merging strings of sequenced data. The power of the merge is equal to T/2.

VRS See "Video Response System."

VSWR See "Voltage Standing Wave Ratio."

VTR See "Video Tape Recorder."

VTR Audio Program Sound A means of recording and playing back audio program material in synchronization with the video signal.

VTR Control Track A control signal recorded onto the tape along with the video information, usually in the form of reference pulses derived from the vertical sync in the signal being recorded.

VTR Pre-Emphasis A means of improving signal-to-noise ratio by increasing the video level at higher frequencies before recording.

VTR Standardization The ability to play back, on the recorder of one manufacturer, tapes recorded on other manufacturer's machines (and vice versa).

VTR Tape Interchangeability The ability to record on one machine and to replay this tape on any one of a number of other machines.

VTR Video Amplification Amplification of the video signal to levels required for recording or modulation.

VTR Writing Speed The relative head-to-tape velocity.

VU See "Volume Unit."

VU-Meter A volume indicator constructed and calibrated to indicate volume in VU.

W-X-Y-Z

WARC See "World Administrative Radio Conference."

WATS See "Wide Area Telecommunications Service."

Wave-Form Monitor A special-purpose oscilloscope which presents a graphic illustration of the video and sync signals, amplitude, and other information used to monitor and adjust baseband video signals.

Waveguide Usually a hollow copper tube of such rectangular or elliptical dimensions that it will propagate electromagnetic waves of a given frequency used for transmitting super-high frequency waves or microwaves. See also "Microwave."

Wavelength Multiplexing Transmitting individual signals simultaneously by using a different wavelength for each signal. Synonymous with frequency division multiplexing.

Weighting Network A network used in or with test equipment for the measurement of noise.

White Clipper A device which prevents the transmission of white peaks exceeding a certain pre-set level.

White Compression Amplitude compression of the signals corresponding to the white regions of the pictures, thus modifying the tonal gradient.

White Level The level of picture signals corresponding to the maximum limit of white peaks.

White Peak The maximum excursion of the picture signal in the white direction during the time of observation.

WIC See "Women in Cable."

Wide Area Telecommunications Service (WATS) WATS permits customers to make (OUTWATS) or receive (IN-WATS) long-distance voice or Dataphone calls and to have them billed on a bulk, rather than individual, per-call basis. The service is provided within selected service areas, or bands, by means of special private-access lines connected to the public telephone network via WATS-equipped central offices. A single access line permits inward or outward service, but not both.

Wideband See "Broadband."

Wideband Channel A channel wider in bandwidth than a voice-grade channel.

Windshield Wiper Effect Onset of overload in multichannel CATV systems caused by cross modulation; the horizontal sync pulses of one or more television channels are superimposed on the desired channel carrier. The visual effect of the interference resembles a diagonal bar wiping through the picture.

Wired City The concept of a fully integrated and featured CATV system providing television, data, educational material, information retrieval, security, utility meter reading and load control.

Women in Cable (WIC) A CATV-industry professional society whose goal is to raise the level of expertise of women in CATV operations, management, and marketing.

Word A basic unit of computer memory. The length of the word may vary from processor to processor. The most common microcomputer word length is 8 bits, or one byte.

Word Processor A computer-based typing and text-editing system.

World Administrative Radio Conference (WARC) A sub-grouping of the International Telecommunication Union (ITU) which is in turn organized by the United Nations for the purpose of coordinating to the extent possible definitions, terms, rules, and uses of the radio spectrum to ensure the highest compatibility while being sensitive to different philosophies and technologies that exist between nations of the world. Conferences are convened every few years and may be either general or specific in focus.

Yagi Antenna A directional antenna array usually consisting of one driven one-half wavelength dipole section, one parasitically excited reflector, and one or more parasitically excited directors mounted in a single plane.

Zero Compression The process that eliminates the storage of insignificant zeros to the left of the most significant digit.

To order additional copies of
JONES DICTIONARY OF
CABLE TELEVISION TERMINOLOGY, 3rd Edition

Send the card below with your name, address, city,
state and zip code along with a check or money
order in the amount of $14.95 plus $1.50 (postage
and handling) for each book to:

JONES DICTIONARY
Jones 21st Century, Inc.
9697 E. Mineral Avenue
Englewood, Colorado 80112

....

JONES DICTIONARY, 3rd Edition
Jones 21st Century, Inc.   9697 E. Mineral Avenue   Englewood, CO 80112

Name _____   Number of Copies _____

Address _____   @ $14.95 each = _____

City _____   Postage & handling

State _____ Zip _____   @ $1.50 each   _____

Total Amount
Enclosed _____

Please allow 4 weeks for delivery.